Editor and Publisher *Paul Mandelstein*
Project Management, Electronic Composition,
  and Technical Illustrations *Professional Book Center*
Design *Lee Ballentine*
Copyediting *Dave Rich*
Proofreading *Lori Kranz*
Indexing *Doug Easton*

Published by Quantum Publishing, Inc.
P.O. Box 310
Mendocino, California 95640

Library of Congress Cataloging-in-Publication Data
    The paging technology handbook / Neil J. Boucher.
      p.   cm.
    Includes index
    ISBN 0-930633-17-2
    1. Radio paging—Handbooks, manuals, etc.   I. Title
  TK6570.M6B69 1992
  621.389'2—dc20
                                        92-44425
                                        CIP

SECOND EDITION, AUGUST 1995

ISBN 0-930633-17-2
Printed in the United States of America

THE PAGING TECHNOLOGY HANDBOOK

Neil J. Boucher

QUANTUM PUBLISHING

CONTENTS

◆ INTRODUCTION

The paging industry has been active for decades, and despite the rapid growth of other mobile technologies, including mobile, cellular, and more recently cordless phones, there is no sign of decline in the demand for paging. Indeed an increasing mobile awareness seems to have bolstered demand for paging, with countries worldwide reporting annual growth rates averaging about 20 percent.

The paging industry has long been one that is highly deregulated, and it is not unusual to find 10 or more operators in one city or region. This competition has been good for the industry in general, but bad for the less sound operators, many of whom have been taken over by larger and better resourced companies.

By its nature, paging is quite labor intensive, and only a few companies have been able to operate solely in the automatic mode. Integrating a large number of switchboard operators with an otherwise high-tech system is not easy. Clearly, full voice recognition encoders are the next major breakthrough for the paging operator, and their arrival will spell the end of an era for the telephone switchboard operator. Voice recognition systems are not yet an economical alternative, but their deployment is probably only a few years away.

The technology available to the paging operators is increasingly complex and requires the operator to master many of the disciplines of modern telecommunications. In the 19 chapters of this book, I have attempted to cover all of the areas of technology that are relevant and necessary for a system of the 1990s.

Neil J. Boucher
November, 1992

CHAPTER

1

 # PAGING—
AN OVERVIEW

Paging has often been regarded as the poor person's mobile communication. It is cheaper than other mobile communications systems because it is a one-way system. The paging receiver alerts the user to the call but does not verify or respond in any way to the base station. Because the cost and bulk of a typical mobile transceiver is associated with the transmit portion (and often the multiplexer—or coupler between the receive and transmit sections), and this is missing from a paging receiver, it can be small and cheap.

In other portable mobile devices such as walkie-talkies and handheld mobile phones, it is typical that half of the weight and bulk is in the battery. This large battery is needed mainly to power the transmitter. Pagers can utilize the smallest of batteries and through intelligent wake/sleep modes can run for months on a single AAA cell.

Early pagers, which used resonant decoders and miniature valves, although bulky by today's standards were at least portable. By the early 1980s, pagers were small enough to be carried with reasonable convenience in the pocket and were comparable in size to the smallest cellular telephones of today. In the next decade they got smaller by an order of magnitude, and the pen-size receiver became a reality. In fact, wrist watches that incorporate a pager and use the strap as an antenna have been on the market for some time.

DTMF

DTMF (Dual-Tone Multifrequency), familiar as the tones on a "touch-tone phone," is now widely available on the Plain Old Telephones (POTs). These tones provide a cheap and convenient means of sending numeric messages from almost anywhere. Paging-system manufacturers were quick to exploit this feature, and most systems include tone access. Some systems use tone access exclusively.

The simplest method of encoding a paging message is to use the DTMF keypad of a telephone to send a numeric message (usually the telephone number to call back). The caller dials the pager required and is then requested to "overdial" the message. This method requires no operators and so is the cheapest to implement. By an extension it is possible to send alpha numeric information with this system using two keys, in turn, to represent an alpha character.

CHANNEL EFFICIENCY

The RF channel efficiency of a digital paging system is very high, and up to 100,000 customers can be accommodated on a single channel. This represents an enormous efficiency in terms of users per channel that is not even approached by other systems. Cellular systems for example may achieve around 20 users per channel. In an area of intensive frequency reuse, the same channel may be reused around 50 times in the same city. At 1000 (20×50) customers per channel, for a high-density cellular system, paging is still more efficient by far. POCSAG operating at 512 bit/s can carry 70,000 tone only plus 70,000 display channels. The bit rate can readily be doubled to 1200 bit/s, which effectively doubles the potential number of users. Trials are under way on bit rates of 2400, while the European ERMES system operates at three times that.

Non-Public Paging

Although most paging systems are public, small private systems for local or regional operation have begun to proliferate recently. Hospitals, restaurants, clubs, and holiday resorts have found in-house paging to be a cheap and simple way to contact staff who may otherwise be difficult to locate.

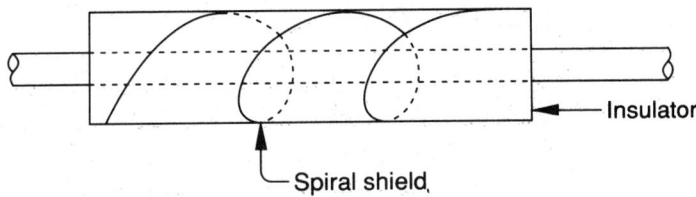

Figure 1.1   *The construction of a lossy coax cable.*

HOSPITALS

Hospitals often have their own in-house paging systems, because their construction makes successful, reliable paging from a public system very difficult. The main problem areas are the radiology and special procedures rooms (which have substantial lead shielding). For most paging companies, the test of building penetration is operation in elevator shafts, but experience shows that even where this is possible it is not uncommon that there are still a number of poor or no-service areas.

It is often said that an in-house system is the ONLY answer to good hospital coverage. Voice messages are often part of the requirement (especially for emergency staff). This type of paging, which for voice requires a S/N of around 18 dB (compared to 0–2 dB for tone only), requires field strengths 15–20 dB higher than for tone only, and therefore exacerbates the problems of reliability and making an in-house system an imperative. Such a system usually comprises a small local controller with a capacity of a few hundred numbers. The best place to mount the antenna is on top of the roof, as this will generally give good building penetration and coverage of the grounds and car parks. In areas that are still poorly covered, the solution is to run a lossy coax cable as shown in Figure 1.1 or to install some localized antennas.

Where it is necessary for a public operator to cover a major hospital, one technique that may be successful is to place two transmitters on either side of the hospital building. This will both improve the penetration and offer a degree of redundancy in the event of a transmitter failure. Considering the known difficulties, however, a detailed survey within the building should be undertaken before any commitments are made.

VOICE MAIL BOX

A recent development is to link a voice message system to the paging network. Voice mail has become so popular that it is now found in the majority of public paging systems. Voice mail enables a caller to leave a specific voice message. The paging customer will be alerted to the fact that the message has been lodged and may retrieve it at a convenient time. Some systems even indicate to the paging customer the number of calls in the voice messaging machine and their nature (that is, priority) usually in three or four categories.

CHARGES

Paging operators make their revenue from billing customers. In the days of tone only paging charging was simple and usually based on a monthly fee. As the paging services got more sophisticated, so charging regimes got more complex.

Charges for paging are often on a fixed monthly basis for tone only calls. This makes billing extremely simple but does reward heavy users at the expense of the occasional users. More sophisticated billing systems allow charges based on usage (generally on top of a fixed monthly charge). Message calls involve an operator, and to recover costs there inevitably will be an airtime cost and often a per-character charge for message calls.

Additional services such as multimode pagers, voice messaging, wide area coverage, call diversion, overhead bulletin services, and roaming mean new sources of revenue and new charges.

MONITORING

A difficulty that is peculiar to the U.S. is an FCC requirement that paging transmitters have a continuous off-air monitor on their transmit frequency which will prevent the transmitter from operating if another carrier is present. While this ensures the peaceful co-existence of a number of systems on the same frequency, in can mean that in the most congested areas delays of up to 10 minutes can occur between a request for a page and the on-air signal. The monitor will be equipped with an activity indicator to indicate the extent of other carriers present.

BACKGROUND

While the cellular industry has been receiving most of the limelight within the mobile radio arena, paging has been growing steadily to a total of 20 million subscribers worldwide in 1990. The total industry investment is comparable to that of cellular and reached $2 billion in 1991.

Although paging is experiencing a healthy growth worldwide, it is particularly active in Southeast Asia, where Hong Kong and Singapore with 120 pagers per 1000 population have the highest penetrations in the world. In both of these countries, it is noted that most cellular-phone subscribers also have a pager. More typical of the developed countries are Australia (35/1000) and Japan (40/1000).

Increasingly, operators are building wide-area (nationwide) systems to meet the demands of their customers for mobility. The world's largest wide-area system is in Taiwan, with 600,000 customers in 1990 and an expansion program to 1,200,000 by 1991. This system is unusual in that it has access only through DTMF overdialing from the PSTN. Table 1.1 (see next page) shows the paging penetration for a number of countries worldwide.

PAGING WORLDWIDE

China

China's paging network was first introduced in 1984. Since then it has undergone a rapid expansion and in 1992 had an estimated 980,000 subscribers. By far the biggest market region is in the industrialized Guangdong area, where there are 200,000 users. The current annual growth rate is around 35 percent.

Hong Kong

Standing second in the world in paging penetration per capita, Hong Kong, with 12 percent and 700,000 users, has impressive paging usage. The majority of mobile phone users also have a pager.

Europe-ERMES

A standard Europe-wide paging system known as ERMES is currently being installed. The forecast of 13 million users by the end of the decade is very bullish from the base of 3 million in 1992. Relatively

Table 1.1   *Paging penetration*

COUNTRY	PENETRATION (%)
Singapore	13.0
Hong Kong	12.0
U.S.A.	5.0
Taiwan	4.5
Japan	4.2
Malaysia	2.7
Australia	2.5
Canada	2.2
Netherlands	2.0
Norway	1.9
Sweden	1.6
UK	1.3
S Korea	1.2
Belgium	1.1
Austria	1.1
Luxembourg	1.0
Denmark	0.9
Finland	0.8
Switzerland	0.5
Germany	0.25
France	0.25
China	0.1
Philippines	0.046

high subscription fees are seen as a disincentive to pager popularity. Agreement is not universal on the value of Europe-wide paging, as most surveys report relatively little Europe-wide roaming by the average user. Although ERMES features enhanced alpha-numeric capability, a resurgence in numeric paging is seen with the increasing accessibility of DTMF telephones.

ERMES operates on a frequency of 169.6 MHz and uses frequency agile pagers. The modulation is 4-level FSK, and the data rate is 6250 bit/s.

Malaysia

Malaysia has a completely deregulated paging market with more than 20 operators. This has not translated to a high usage rate as the subscriber population is only 50,000, making it one of the smallest in the region.

Singapore

Singapore has the highest penetration of pagers in the world. This island city-state is the business hub of Southeast Asia. The paging networks are run by Singapore Telecoms, which is a government-owned body (but is currently on the market for privatization). There are 8 numeric, 1 alpha, and 1 international 931.9375 MHz systems for a subscriber population of 450,000. Most of the pagers (93 percent) are numeric. The system uses POCSAG in the frequency range 150–170 MHz on 21 channels. The busy-hour calling rate is a high 0.8 calls/subscriber. This high calling rate means that only about 20,000 users can be supported on the 512 bit/s system.

Singapore has long had a policy of development based on good, efficient, and affordable telecommunications. Both the terminal equipment cost and the monthly fees are kept to a minimum. The flat monthly fee, with no air-time charge, is $18 SIN (about $11 U.S.). It is perhaps the affordability, more than any other factor, which accounts for the 13 percent penetration and the growth rate of 6000 subscribers per month.

Philippines

There are two operators in the Philippines and 30,000 pagers. The market has been expanding slowly and is concentrated in Manila.

Taiwan

The Taiwan paging system was installed as a large-scale, economically priced, tone-only network. It has rapidly grown to have one of the highest penetrations in the world, with well over 1 million subscribers in the largest single wide-area network in the world.

BASIC RADIO

Radio was first postulated in 1873 by Maxwell, demonstrated in 1888 by Hertz, and used for practical communications in 1895 by Marconi. Radio is an electromagnetic phenomenon and radiates as photons. It belongs to the family of radiation that includes X-rays, light, and infrared (heat) waves. The different categories of radiation differ in frequency, as shown in Figure 2.1. They also differ in energy and ability to propagate those different media.

BASIC ELEMENTS

All practical radio systems can be reduced to the basic scheme shown in Figure 2.2.

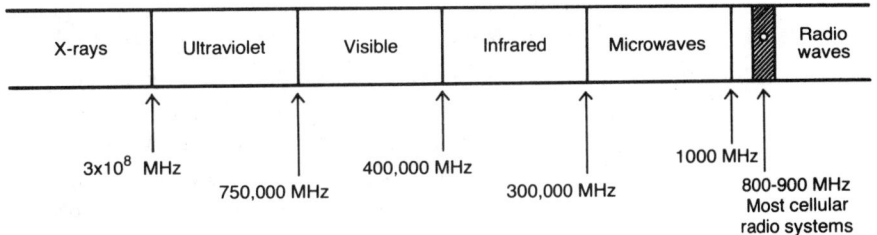

| X-rays | Ultraviolet | Visible | Infrared | Microwaves | | Radio waves |

3×10^8 MHz   400,000 MHz   1000 MHz

750,000 MHz   300,000 MHz   800-900 MHz
Most cellular
radio systems

Figure 2.1   *Electromagnetic spectrum.*

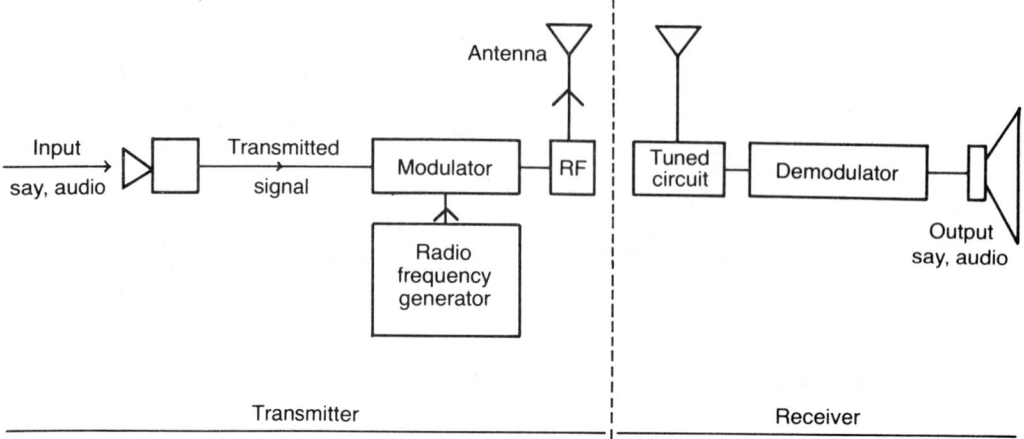

Figure 2.2   *A simplified analog transmitter/receiver with an audio AM modulation.*

Transmitter

The transmitter consists of two basic parts: a modulator and a carrier. Figure 2.2 shows an analog transmitter with an audio input that is converted into the form to be transmitted (in this case, via a microphone to the modulator). A radio frequency generator generates the radio energy that will carry the signal. This generally consists of an oscillator (which produces the initial signal) and a number of amplifier stages (which amplify the level to that required at the antenna). A modulator mixes the signal to be transmitted with the radio frequency signal (called the carrier) in such a way that the signal can be decoded at a distant receiver.

Receiver

The receiver in Figure 2.2 gets a signal from its antenna, which also receives a number of unwanted signals. The tuned circuit tunes out all but the wanted signal, which is then demodulated (decoded) by the demodulator.

The very simple receiver illustrated in Figure 2.2, consisting only of a tuned section and a demodulator, is known as a *tuned radio frequency* (TRF) receiver. Until about 1930 most receivers were of this type, and they all suffered from a lack of selectivity.

The TRF was replaced as the main commercial receiver by the *superheterodyne*, commonly known today as the *superhet* and shown as Figure 2.3. The name was derived from the word *heterodyne*, which means the beating of two signals together to produce the sum and dif-

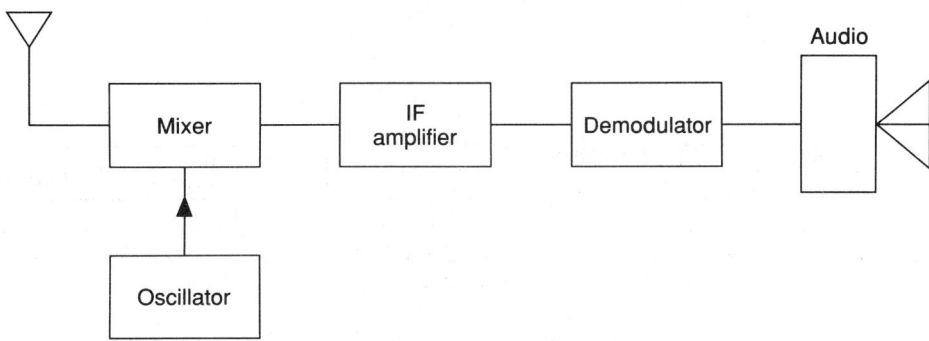

Figure 2.3   *A basic superheterodyne receiver.*

ference frequencies; (CW or *continuous wave* Morse code signals were decoded by beating the incoming signal with one of almost the same frequency so that an audio tone equal to the difference frequency was heard). The tuning and signal amplification in the early superheterodynes was done at supersonic frequencies of around 50 to 60 kHz (the heterodyne frequency), and hence the word "super." The heterodyne frequency is now known as the *intermediate frequency* or IF, and in modern receivers is well above the supersonic range—typically 455 kHz or 10.7 MHz.

An extension of the superhet principle is the *double conversion superhet*, which is commonly used in most cellular receivers, as well as in other high-frequency domestic sets. This receiver uses two IF stages in series, the first being typically at 10.7 MHz and the second at 455 kHz. This will provide superior RF gain with stability and virtually eliminate the main shortcoming of the superhet, which is its image rejection. The heterodyne action will produce both the sum and difference of the incoming frequency and the oscillator. If the incoming frequency was, for example, 800 MHz and only the 455 kHz IF was used, a mixer operating at 800.455 MHz will result in the required IF signal at 455 kHz. However when the receiver is tuned for an incoming signal of 799.090, the oscillator frequency would be 799.545 MHz (800 MHz–455 kHz), both the the original 800-MHz signal, and the desired 799.545-MHz signal will produce a beat frequency at 455 kHz. The 800-MHz signal will do so as a result of the difference frequency between it and the oscillator. In other words, two oscillator frequencies very close together will produce a valid signal to the IF amplifier. This unwanted signal is known as an *image frequency*, because unless the RF filtering screens the unwanted signal completely, the tuning will result in there being two positions on the tuning dial at which the

800-MHz signal is found; the undesired position is known as an *image*.

Clearly it would be difficult to provide adequate RF tuning at 800 MHz to ensure a perfect filtering of signals only 910 kHz (2×455 kHz) apart and for this reason the first IF stage frequency will be higher; typically 10.7 MHz. Now, the same problem will still occur, but the frequency at which the receiver will be tuned to generate the image of an 800-MHz signal will 778.6-MHz (800 to 2×10.7 MHz) and the RF tuning needs only to be able to screen out the 800-MHz signal, which is 21.4 MHz away.

Additional stability can be provided by utilizing a second IF frequency, as it can be seen that a high-gain tuned radio frequency stage is a potential oscillator if the leakage from input to output becomes sufficient. Usually that second stage will operate at 455 kHz, since the 910 kHz margin to the image frequency is quite manageable at the down-converted frequency of 10.7 MHz.

The choice of the actual IF frequencies is limited to those that have by convention been set aside for this purpose. No transmitters operate on these frequencies and so they will be noise-free.

A final improvement that is characteristic of modern receivers is the addition of an RF stage, which consists usually of a single stage, low-noise amplifier to improve the signal-to-noise performance of the set. In fact, it will be seen later in the chapter on noise performance that in a normally operating superhet it is the noise figure of that RF stage that will virtually determine the overall receiver performance. These elements will be found in a typical paging receiver, shown in Figure 2.4.

Modulator

The modulation system used in 2 and 5 tone paging receivers is known as FM (Frequency Modulation). In this type of modulation the frequency of the carrier is varied proportionally to the signal to be transmitted. A typical FM modulator is shown in Figure 2.5.

The audio input varies the bias on the varactor (a solid state variable capacitor, illustrated in Figure 2.5), which in turn changes the frequency of the tuned circuit. The maximum amount that the frequency can deviate from its central carrier frequency is called the peak deviation.

The S/N performance of FM systems is very high, provided the noise level is reasonably low. FM systems with wide deviation have better S/N performance than those with narrow deviation.

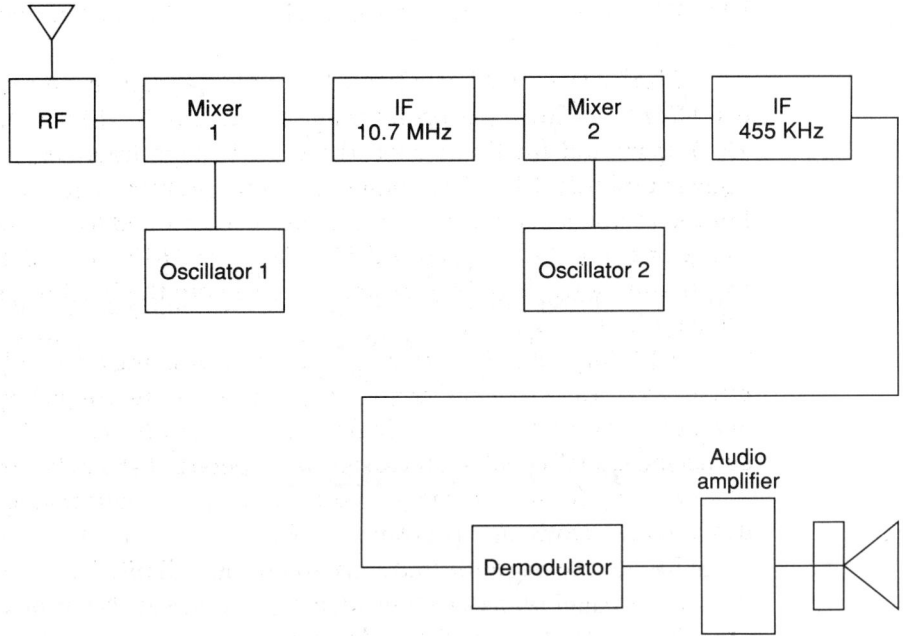

Figure 2.4   *A typical superhet configuration of the type used in cellular radios.*

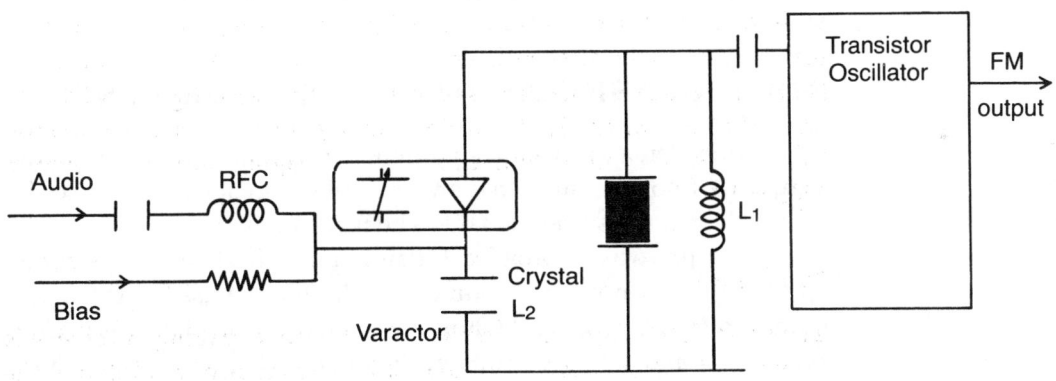

Figure 2.5   *Typical FM modulator.*

Some systems use phase modulation, particularly for data transmission. Phase modulation is closely related to frequency modulation and can be derived from it by passing the signal through a simple differential circuit before frequency modulation (see Figure 2.6).

Frequency Shift Keying (FSK), at relatively low speeds, is often used for data, because it has better S/N performance than FM at low signal levels. This enhances signaling in areas of poor reception.

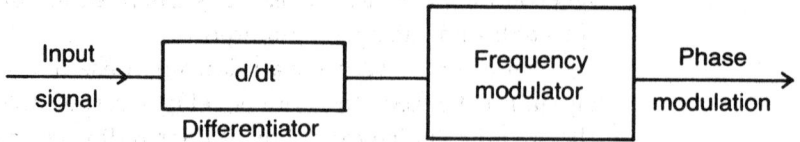

Figure 2.6   *Phase modulation from frequency modulation.*

As the signal level increases, the quality of FM (in noise performance terms) rises fairly rapidly, whereas the quality of FSK does not. Because cellular systems are designed to operate in relatively high signal environments (low noise), FM is chosen for the voice path.

DYNAMIC CHANNEL ALLOCATION

Because many modern paging systems use multiple channels, it is necessary for the pagers to automatically switch to the correct channel in use. The pager scans and then switches to the channel indicated by using synthesized tuning, a system where the frequency of the oscillator is numerically compared to the required frequency and adjusted by a "phase-locked loop" until the two frequencies match.

NOISE AND SIGNAL-TO-NOISE PERFORMANCE

All radio systems are ultimately limited in range by noise. When the intrusion of noise is such that an acceptable signal can no longer be obtained, then the system is said to be noise-limited.

Medium wave (broadcast band) and shortwave broadcasts operate in a very noisy environment; background noise limits the performance in the broadcast/shortwave bands. The VHF (Very High Frequency) and UHF (Ultra High Frequency) bands where cellular radio operates are relatively much quieter and most of the noise is generated in the radio frequency preamplifier of the receiver itself. Regardless of how well designed the receiver is, there is a theoretical noise power level which, at a given temperature, cannot be improved upon. This is because of the thermal noise, generated by the movement of atomic particles (in the receiver and most particularly in the first radio frequency amplifier). This noise is proportional to the operating temperature. Hence, the antenna and RF amplifier stages will generate thermal noise continuously. For this reason high-quality

receivers, such as radio telescopes, operate their input stage RF amplifiers at liquid-nitrogen temperatures.

In order to perceive a relatively noise-free signal, the incoming
signal must exceed the noise level by a respectable margin, known as
the signal-to-noise ratio. For cellular radio systems, this level is usually regarded as 12 dB for marginal reception and 30 dB for good quality conversations.

Signal-to-noise ratio is usually expressed as:

$$S/N = \frac{\text{Signal level}}{\text{Noise level}} \text{ (usually expressed in dB)}$$

Because modern receivers closely approach the theoretical noise
limits for their operating temperatures, it can easily be deduced what
minimum received signal level is required to achieve a satisfactory
signal-to-noise ratio. Pagers are designed to operate at signal-to-noise
levels of a few dB and so require elaborate error correction codes.

dBs

Humans perceive power logarithmically. For example, doubling the
energy level of a sound pulse produces only a 3 dB increase in the per-
ceived level—and that increase is only just noticeable. The term dB
was introduced to define relative power levels logarithmically.

The term dB is used often in radio systems and can be a major
source of confusion to the uninitiated because of the large number of
different units of dBs. Essentially, the dB level is the log of a power
ratio: dBm, the most common form of dB, is the power of the system
measured compared to 1 milliwatt. Mathematically, this can be
expressed as:

$$\text{Power dBm} = 10 \log \left[\frac{\text{Power (in watts)}}{(0.001)} \right]$$

$$\text{Power dBm} = 10 \log \left[\frac{\text{Power (in milliwatts)}}{1} \right]$$

$$\text{Thus 1 watt} = 10 \log \frac{1}{0.001} \text{ dBm} = 30 \text{ dBm}$$

dBμV/m is a unit of field strength which compares the measured
level with 1 μV/m (1 microvolt per meter).

Mathematically, this is:

$$dB\mu V/m = 20 \log \left[\frac{\text{field strength in microvolts per meter}}{1} \right]$$

NOTE 20 is the multiplying factor here because the terms being used are voltage, not power. Voltage squared gives the power ratio.

PROPAGATION

Radio propagates at the speed of light (299,800 km/sec, or approximately 300,000 km/sec). Medium- and high-frequency waves can propagate very long distances by reflecting off the ionosphere, as shown in Figure 2.7. This is the way shortwave propagates around the world.

At higher frequencies (above about 50 MHz), the troposphere/ionosphere absorbs the waves instead of reflecting them so that the predominant mode is the direct wave. The direct wave is not limited to line of sight; in fact, a good deal of refraction (bending of the path of propagation) occurs. This enables the transmissions to extend well beyond line of sight. Diffraction (bending around obstacles) also occurs, allowing the path of the wave to extend around obstacles. The ability to refract and diffract decreases with increasing frequency but is still most significant at the higher paging frequencies (800–900 MHz).

A third property, reflection, is also significant at paging frequencies. Coverage in high-density city areas is significantly enhanced by

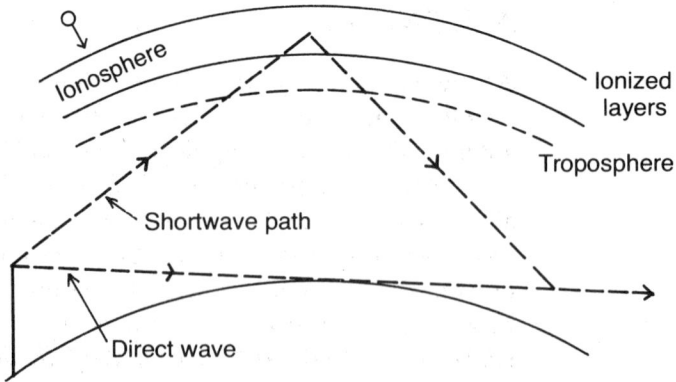

Figure 2.7   *Waves reflecting off the ionosphere.*

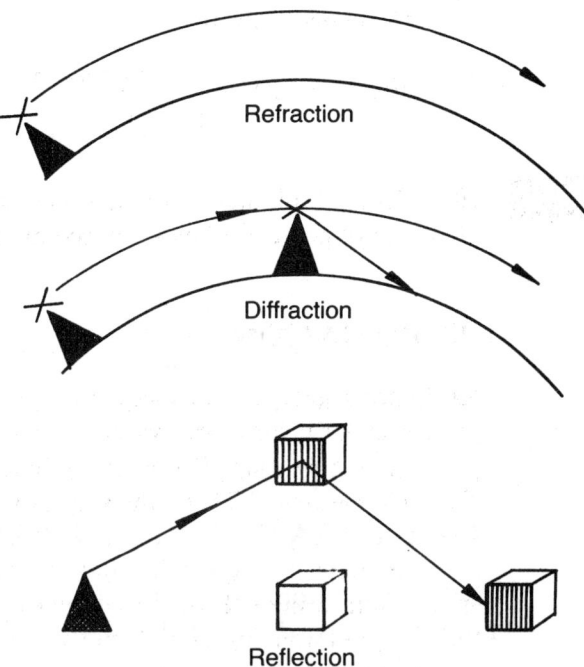

Figure 2.8   *Modes of propagation at cellular frequencies. Super-refraction, or ducting, is a sporadic phenomenon responsible for propagation over very large distances under certain atmospheric conditions.*

the ability of the radio system to reflect into most areas that are inaccessible via a direct path.

These three modes of propagation are summarized in Figure 2.8.

ANTENNAS

Antennas used in paging base stations are usually gain antennas, meaning that they have gain, compared to the simplest form of antenna, the dipole. A dipole is shown in Figure 2.9.

Mobile paging antennas known as "beeper boosters" are usually between 3 and 4.5 dB gain. Base station omnidirectional antennas, which stack many radiating elements in series, are often 6 to 9 dB in gain. Very high gain antennas are only practical in fixed locations, because they are very large and must be exactly vertical to operate satisfactorily. Sometimes, however, they are deliberately tilted down to limit the base-station range.

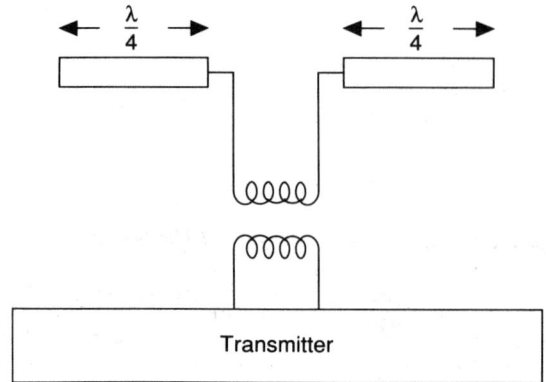

Figure 2.9   *Dipole antenna.*

MOBILE TRANSMIT POWER AND HEALTH

The radiation from a radio system is non-ionizing (distinguishing it from radioactive decay products); the main effect on the human body is a rise in temperature. Contrary to popular belief, it is not the high mobile frequencies that are inherently most harmful to humans. At the lower frequencies—around 100 MHz (often used for paging)—the body can become resonant and, therefore, very absorbent. Hence these lower frequencies are potentially more hazardous. Early experiments using small laboratory animals pointed to relatively more harmful effects at higher frequencies. However, the small size of the animals, which gave the animals a high resonant frequency, probably accounted for these results.

Care should be exercised when working near paging base-station antennas, particularly those operating at high ERPs. Where it is necessary to work closer than 1 meter away or for periods greater than 10 minutes at a time, safety procedures should be followed.

CHAPTER

3

 # PAGING RECEIVERS

A paging receiver typically has a single loop antenna, which enables it to operate at around 6 µV/m in free space or 10 µV/m on the body.

Paging receivers are usually double conversion superheterodyne receivers, as seen in Figure 3.1. The performance of a paging receiver is highly dependent on its sensitivity, and as the antenna is not very efficient (typically having a gain of –25 dB, referenced to a dipole) it will always have an RF amplifier, as seen in the figure, with a noise figure of about 4 dB. The double conversion IF will contribute to improved selectivity and will consist of a first IF of 10.7 or 21.4 MHz and a second IF of 455 kHz or 30 kHz. Some, however, are direct conversion, with most of the filtering done at audio frequencies.

BATTERY SAVING

Most of the approximately 5 mW of battery consumption of a modern pager is dissipated in the RF and audio circuitry. The decoder, which is CMOS, consumes only about 200 µW, and so battery saving schemes have been implemented whereby the pager RF section can be switched off for the duration of messages not relevant to it. Most pagers are powered by a single AAA 1.5 V cell.

The preamble in a 512 bit/s POCSAG system will last 1.125 seconds. This means that if the paging receiver activates itself for a

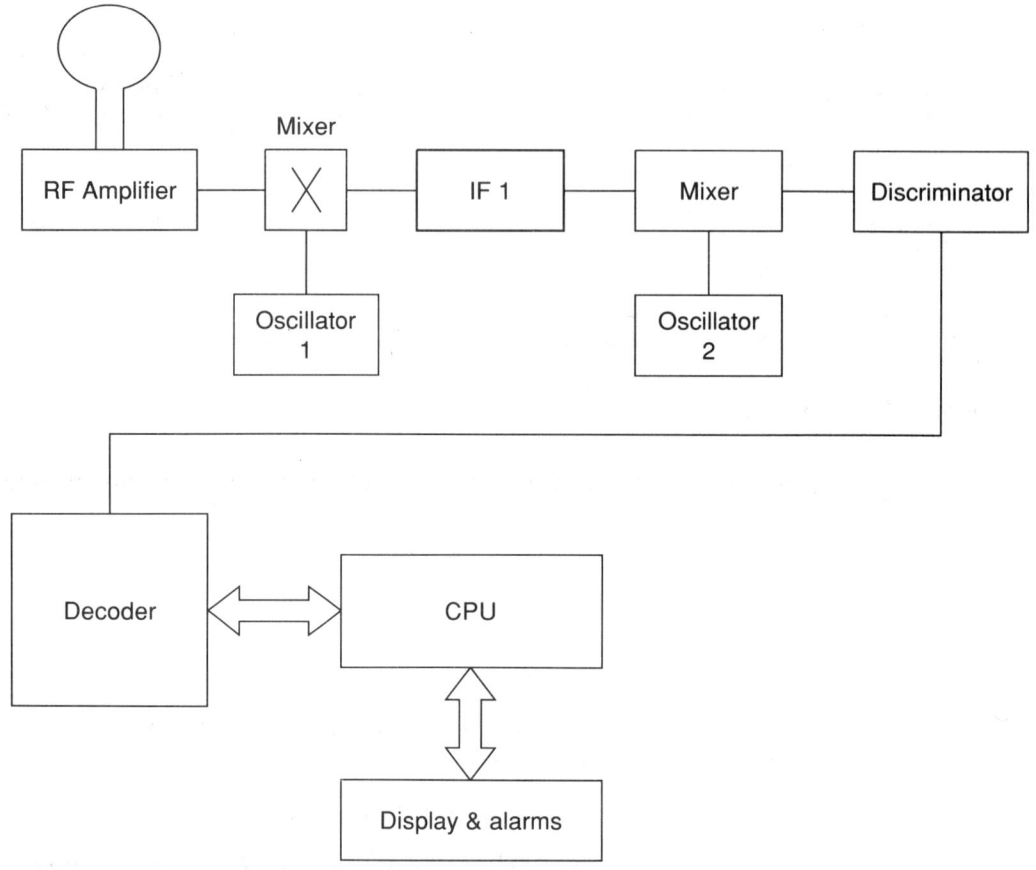

Figure 3.1   *The basic structure of a paging receiver.*

period shorter than 1.125 seconds, but long enough to detect the pre-amble sequence, then in the event of no page, it can power down for the rest of the time without missing a page. Typically a paging receiver may turn on once every second for 70 milliseconds. In that 70 milliseconds it will determine if it is a preamble period, and if so will continue to listen. If not it will shut down for another 1 second. Assuming a period of zero calls, this will amount to a duty cycle of 70/1000 or 0.07. This process is illustrated in Figure 3.2.

THE PAGER ANTENNA APERTURE

The concept of the effective aperture of an antenna has long been used as a means of comparing the efficiency with which different antennas extract energy from an electromagnetic field. The horn antenna, as depicted in Figure 3.3, shows the antenna capturing the energy of an

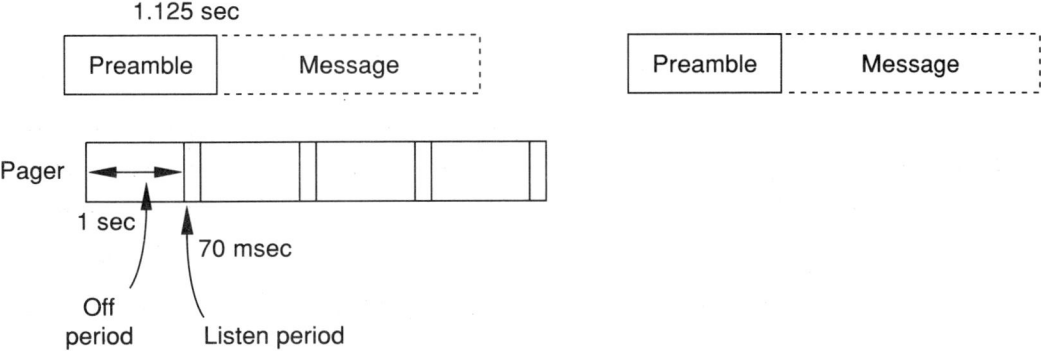

Figure 3.2 *A wake/sleep sequence for a pager.*

incoming wave. It can readily be seen that for a 100 percent efficient horn antenna the power captured will be:

$$P = S \times A$$

where

$P =$ the power captured in watts

$S =$ the power density of the electromagnetic wave
(measured in watts/m^2)

$A =$ the area of the horn mouth in m^2

For other antennas it is a little more difficult to visualize what is meant by aperture. Consider the case of a half-wave dipole, which has an effective aperture of 0.13 × (wavelength)^2. It can easily be visualized that in the far field there will be a uniform RF power density S (known as the Poynting vector). For an antenna to be effective, it must capture RF over an area in a manner similar to the horn antenna. This can be visualized as an effective capture area extending along the length of the antenna and to a depth of a quarter wavelength, as depicted in Figure 3.4(a).

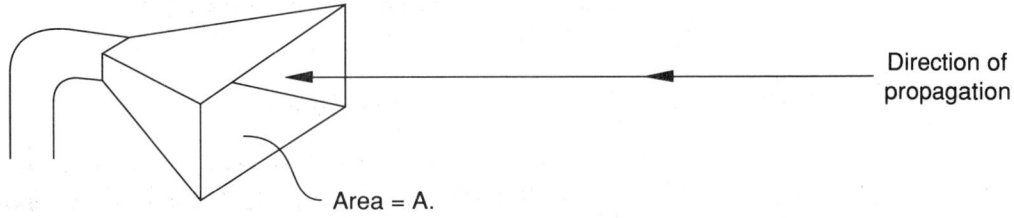

Figure 3.3 *A plane wave incident to a horn antenna.*

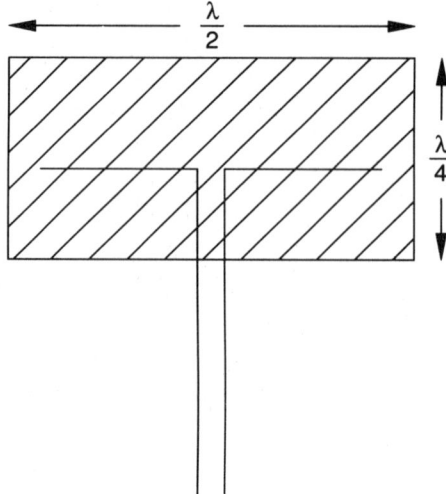

Figure 3.4(a)   *The effective aperture of a dipole antenna.*

Finally, consider the aperture of a small loop antenna, which can be shown to have an aperture of $0.119 \times$ (wavelength)^2. The essential point to note here is that the aperture is *independent* of the loop size (provided the loop is small relative to a wavelength) and is proportional to the square of the wavelength, as seen in Figure 3.4(b).

RADIATION RESISTANCE

An antenna will have a characteristic impedance which can be determined by the relationship between the induced voltage and currents in it. For a lossless antenna with an aperture A_e, the received power will be $A_e \times S$. The equivalent circuit of the antenna and load is shown in Figure 3.5. By measuring the induced voltage into a matched load and assuming a 100 percent energy transfer, the following relationship is obtained:

$$A_e \times S = (V/2)^2 / R_r$$

where

R_r = the radiation resistance.

The radiation resistance is fictitious and should not be confused with a real resistance.

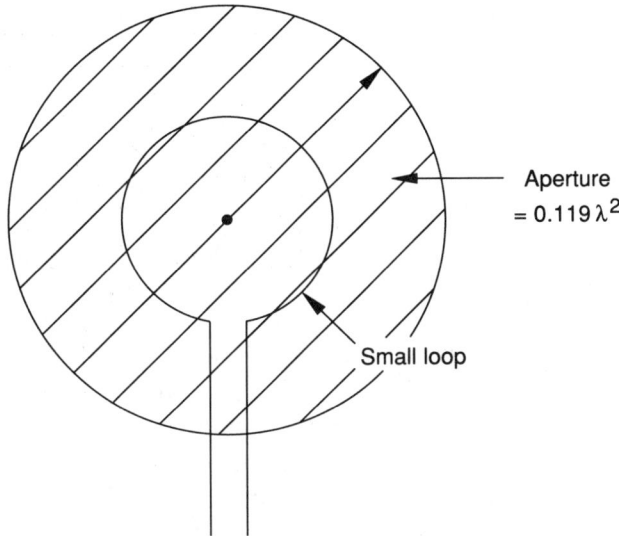

Figure 3.4(b) *The aperture of a small loop antenna.*

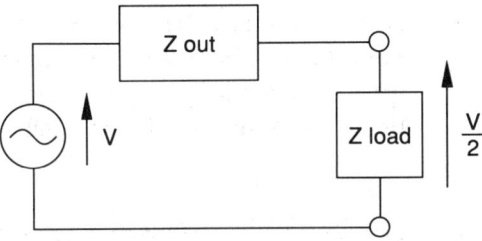

Figure 3.5 *The equivalent circuit of the antenna and its load.*

Where the antenna has a significant real resistance (RR) this will reduce the aperture by a factor of $R_r / (R_r + R_l)$.

PAGER SENSITIVITY AND RF DESIGN LEVEL

Commercial sensitivity specifications for pagers usually define the field strength, in microvolts per meter, at which the pager receives calls in a steady RF field with a 50 percent probability while the pager is randomly oriented around the vertical plane. One way of doing this is to place the pager in an antenna range and orient it so that the loop is progressively rotated through 360 degrees and record the call success rate at various field strengths. Often the specification will involve the measurement being done while the pager is being worn by a person.

This introduces an uncertainty in the measurements, as while it is true that the sensitivity will be reduced by the presence of a person, it is also true that there is no such thing as a standard person (so the body mass and height of the wearer may affect the results).

Using these standards, pagers will be found to be rated at sensitivities varying from around 5 to 50 µV/m. As an example, consider a pager that is rated at 6 µV/m operating on 150 MHz. As a system designer, there can be no doubt that a 50 percent fail rate is totally unacceptable, and that a much higher success rate would be desired. Also, in the real world the pager will not be operating in a steady RF field, but in a much more hostile environment. To determine a design field strength, first convert the field strength to a more manageable logarithmic form: 6 µV/m = 20 × Log 6 or 15.6 dBµV/m. Experiments show that the difference between a 50 percent fail rate and a 90 percent fail rate is about 4 dB and that an allowance of 10 dB is needed to maintain the same bit error rate in the multipath environment as in a steady field. This leads to a design field strength of 29.6 (or nearly 30) dBµV/m.

Alpha-numeric pagers are more sensitive to error than alert only units, as even a single error in the message will be noted by the subscriber. An extra allowance in field strength of around 4 dB should be made in this case.

THE PAGING RECEIVER ANTENNA

A paging receiver will generally have a simple built-in loop antenna (as depicted in Figure 3.6), which because of its size will be inefficient. Because the loop is small relative to a wavelength, the current distribution in the loop can be considered to be almost uniform, and the antenna behaves very much the same as a short dipole. The characteristics of the loop antenna are summarized in Table 3.1 together with the corresponding parameters for a dipole and a short dipole.

It is of interest to note that the aperture of a small loop is independent of the loop size and proportional to the wavelength squared. Also it should be noted that the gain of the loop relative to a half-wave dipole is constant with frequency variations. At first sight this may lead to the impression that the longer the wavelength the more effective is the loop. This, however, is not the whole story as can be seen when the radiation resistance is considered. As an example, assume

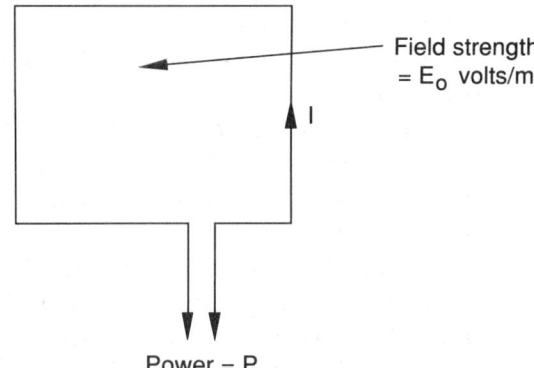

Figure 3.6   *The construction of a simple paging antenna.*

Table 3.1   *Radiation resistance and effective aperture of a small loop, a short dipole, and a half-wave dipole.*

Antenna type	Radiation resistance	Effective aperture*
Short dipole	$80 \times (\pi \times L/(\text{wavelength} \times \sqrt{2}))$	$0.119\ \lambda^2$
Half- wave dipole	73	$0.13\ \lambda^2$
Small loop	$31200 \times (A/(\text{wavelength})^2)^2$	$0.119\ \lambda^2$

that the pager operates on 100 MHz and has a loop with the dimensions 6cm × 4cm (0.0024 m^2), and a wavelength of 300/100 or 3 m. The lossless radiation resistance will be an incredibly small 0.00222 ohms. At 1000 MHz this radiation resistance becomes a much more manageable 22.18 ohms, as depicted graphically in Figure 3.7.

The importance of the radiation resistance can be seen when the effect of the ohmic losses in the antenna and the front end of the paging receiver are considered. It is very difficult to match very low impedances (of the order of milliohms) into the preamplifier of the paging receiver without incurring significant losses.

It can be seen that the gain of the loop is very dependent on frequency and will increase by 4 times (6 dB) for each doubling of wavelength. For the case where the loop and receiver impedances are matched this relationship is as shown in Figure 3.8.

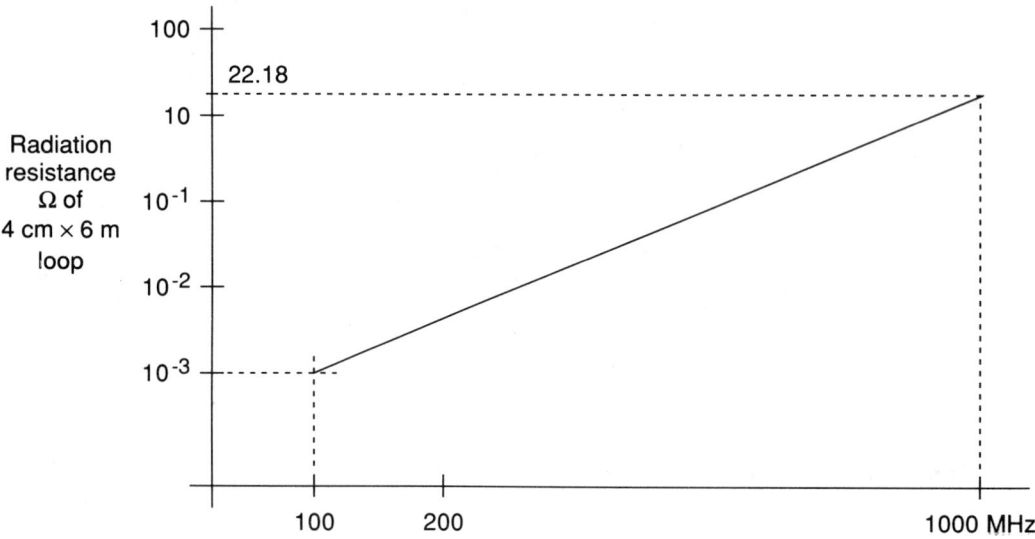

Figure 3.7   *The radiation resistance as a function of frequency for a 6cm × 4cm loop antenna.*

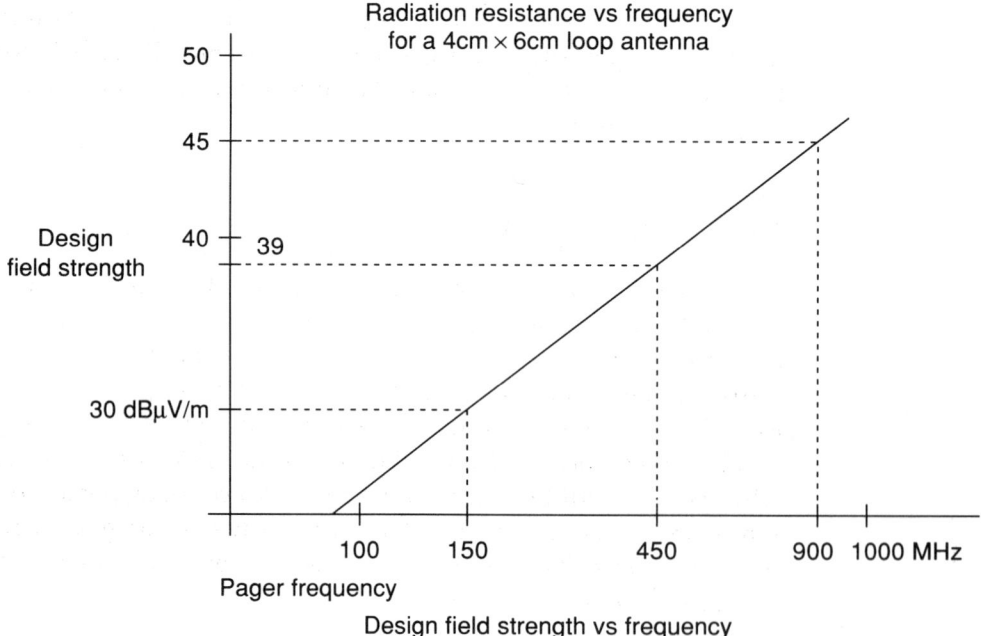

Figure 3.8   *The frequency dependency of loop gain, assuming loop and input impedances match.*

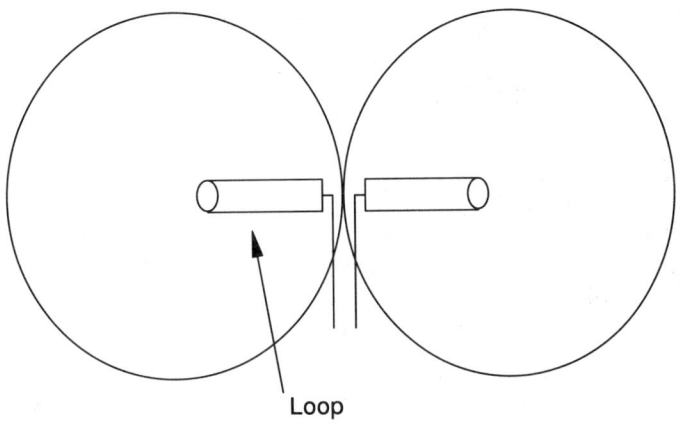

Loop

Figure 3.9   *The far-field pattern of a small loop antenna.*

DIRECTIVITY

The directivity of an antenna is defined as the ratio of the maximum radiation intensity compared to the average intensity. For a small loop this is 1.5, which is the same as the directivity of a small dipole and similar to a half-wave dipole (which is 1.64). The pager is, however, used at all angles, and in practice the gain in the null direction is about 6 dB down from the maximum. The directivity of a small dipole is depicted in Figure 3.9.

BEEPER EXTENDERS

One of the factors that most restricts the range of a pager is the inefficient antenna, which is a natural consequence of the small size of a modern receiver. In some rural areas, where the paging range is restricted, additional coverage can be obtained by the use of a "beeper extender." This device consists of an antenna mounted on a vehicle window, together with a small bracket on the inside of the window which serves both to hold the pager and as a coupling device to the external antenna. It works much the same way as an on-glass mount, cellular antenna. It can improve the effective sensitivity of a pager by as much as 10 dB.

PAGING REPAIR

In the early 1980s paging was almost entirely analog, and the pagers were constructed of relatively large components, many of which were discrete. Today the units are constructed of surface mounted CMOS ICs, which are both difficult to service and very vulnerable to electrostatic damage.

As the prices of pagers have been decreasing over the years (even in dollar terms), so the tendency has increased to regard the units as non-repairable or at least only factory repairable. The operator will find that it will be economic to undertake at least minimal repairs on site as a surprising number of faults turn out to be as simple as dead batteries, dirty battery contacts, or even that the pager is not switched on. Without any facilities at all these problems can be checked and a test call made to confirm operation. With the extended warranties being offered by some manufacturers today, this repair level may be adequate for most small operators. However, consideration must be given to the delays that can be expected when the pager is sent off for service, which will typically be several months. Some arrangements to keep the customer on the air in the meantime will need to be made. This usually means loaning the customer a pager.

Should the operator wish to undertake further repairs, then some equipment and training will be needed for board level (meaning repair by replacement of boards or other assembled parts of the pager). A trained technician should be able to capably undertake this level of repair with specialized training (usually provided by the manufacturer) amounting to a few days, but in special cases extending to one or two weeks. The equipment needed will include a work bench with good ground mats and wrist straps, a selection of spare boards, a temperature controlled soldering station, multimeters, an oscilloscope, and service tools. A budget of around $3,000 per service position should be adequate for tools, plus a parts budget, which would depend on the demand.

Paging repair has become very much a specialist function, done mainly by the supplier or a centralized repair station. In order to effectively undertake repair, it will be necessary to have at least one well equipped, shielded room to enable sensitivity and other RF tests to be carried out in a relatively RF quiet environment.

FAULTS

High impedance CMOS devices are used to reduce the current drain and so increase the battery life. Surface mount is used to improve the MTBF and to decrease the physical size of the unit. These same features make the unit very susceptible to failure due to even very small amounts of water or dust. Once lodged, the contaminant may be difficult to see with the naked eye and will likely cause intermittent faults. Inspection under a bench magnifier may reveal the problem. Pagers are generally not water resistant and even the conformal coatings manufacturers have used to limit water damage are not always complete and are also very prone to rupture.

Removing corrosion, if it is not far advanced, can be done with a cleaning agent like methylene chloride but will require total disassembly of the unit as the agent can cause damage to plastic and rubber components. It should also be remembered that these cleaning agents represent a health hazard and should only be used where there is adequate ventilation.

Sensitivity tests, which can effectively be done only in a shielded room, are needed to ensure the correct functioning of the RF section. SINAD tests are done by injecting a carrier of a known level and deviation and measuring the resultant received signal-to-noise ratio, as depicted in Figure 3.10.

A sensitivity of around 0.35 µV for 12 dB SINAD should be expected. Some technicians are in the habit of just listening to the "noise level" instead of using a SINAD meter. Although this may work sometimes (or even often) it is not a good practice and should be discouraged. The ear may be able to estimate to a few dB, but in fringe areas, particularly with digital pagers, a few dB can mean the difference between working and not working.

SINAD measurement is one of the most fundamental measurements; its method is best illustrated using a separate RF generator and SINAD meter. Figure 3.11 shows a SINAD meter. The RF generator sends a signal at a specified test frequency and deviation (usually about 1-kHz frequency and 1-kHz deviation) to the receiver. The SINAD meter has internal filters that separate the test tone (wanted signal) from the other components (assumed to be noise and distortion from the receiver).

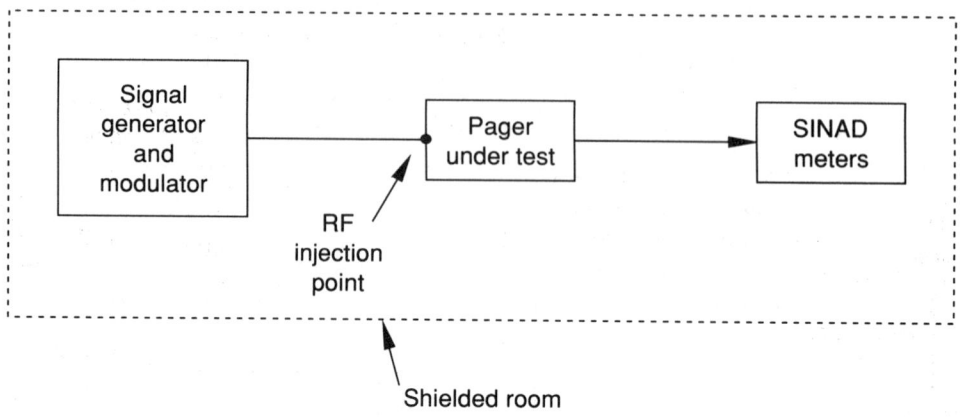

Figure 3.10   *The measurement of SINAD.*

A meter reads the SINAD directly and the RF level is adjusted until the desired SINAD is obtained. Various weighting filters are sometimes used to limit the bandwidth and pass characteristics of the noise and distortion components. This is because the receiver channel itself is filtered and the receiver transducer (earpiece or decoder) and the ear combined do not respond to all frequencies equally. Thus, the "weighting" networks attempt to account for the subjective noise-level rather than the actual S/N levels.

The distortion function of the SINAD meter can be used in this same test to measure the receiver distortion. If the RF signal level is high enough the processing gain of the receiver will reduce the noise to around 70 dB below the signal, and at this level it is negligible. If the 1-kHz tone level is now raised so that the signal generator is at 25 percent of the maximum system deviation, it will be at an energy level that corresponds to the average of a good signal. The SINAD meter will now be reading the level of distortion in dB relative to the signal. This should be below 5 percent (although this specification may vary a little from system to system) or 10 log 0.05 = -13 dB relative to the tone.

Although only one service monitor is needed for in-band tests, at least two are needed for out-of-band testing; the second monitor can also be used as a reference (that is, discrepancies between monitors will alert the user to a calibration error). As frequency congestion increases, the possibility of adjacent-channel interference increases and the need for out-of-band measurements increases.

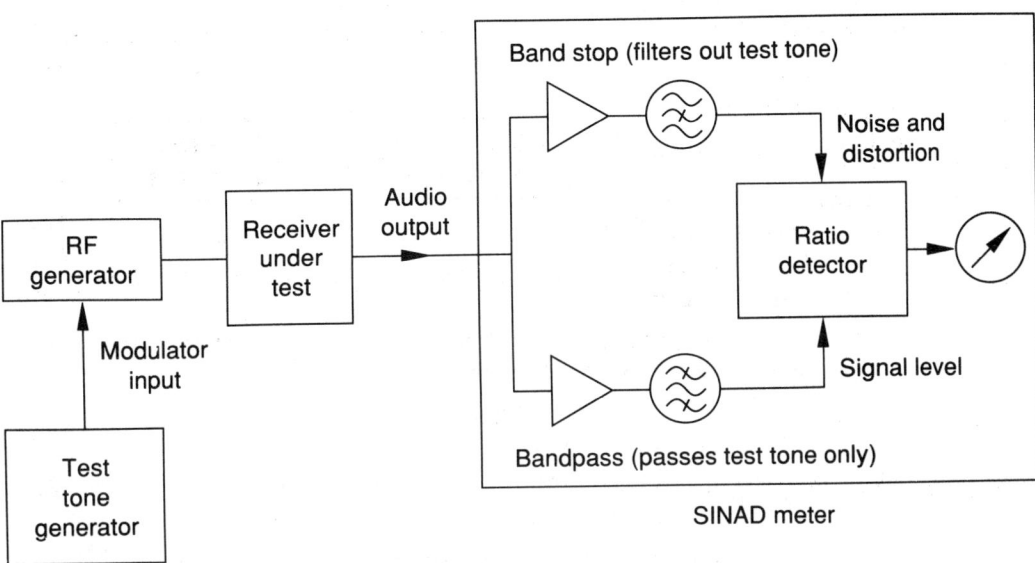

Figure 3.11   *A SINAD meter.*

Measuring adjacent-channel rejection requires that one signal generator be set to measure 12-dB SINAD in-band, and the second signal generated be coupled to the receiver input and set to the adjacent channel. The second generator is adjusted in level until the SINAD falls by 6 dB, as shown in Figure 3.12.

In general, it is a good policy to have at least two of each test equipment item so that the calibration of one can be checked against the other. Downtime with test equipment can be excessive (repair

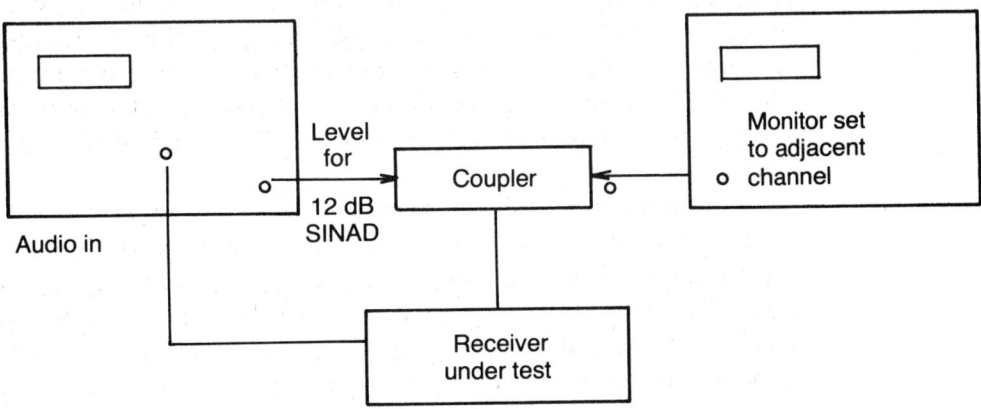

Figure 13.12   *Two signal generators are needed to measure adjacent channel rejection.*

times of 3–6 months are not unusual), and the second piece of equipment can be extremely valuable at this time.

For digital pagers to work properly, the decoder, which will have an important part to play in correctly receiving the signal in a noisy environment, should also be tested. This is done usually by generating a number of pages, sent at a level near the threshold and ensuring that they are correctly received. Naturally this will mean that a signal generator that can imitate the paging transmitter code will be required.

Where multiple boards are used, the connectors between them are a frequent source of problems.

 UNITS AND CONCEPTS OF FIELD STRENGTH

There are a diversity of units of field strength in use in RF engineering today. Mobile engineers tend to prefer the unit dBμV/m, but some still use units like dBm, which have their origins in the land-line network. This chapter seeks to clarify the usage of the different units and the concepts behind the measurement of field strength.

In free space, the energy of an electromagnetic wave propagates through space, and, according to an inverse-square law, with distance. The energy is dispersed over the surface of an ever-increasing sphere, the area of which is R (where R is the sphere radius). The total energy is constant because there is no loss in free space. The energy measured in watts/square meter (or any other units) will be constant in total and so the energy per unit area will vary as $1/R^2$. Notice that this equation holds for all frequencies. This concept is illustrated in Figure 4.1.

The total energy within a solid angle is a constant at any radial distance from the origin.

In a paging or mobile environment, the signal is attenuated much more rapidly than in free space and follows approximately an inverse fourth power law with distance. There is a common misconception that this attenuation increases rapidly with frequency. It will

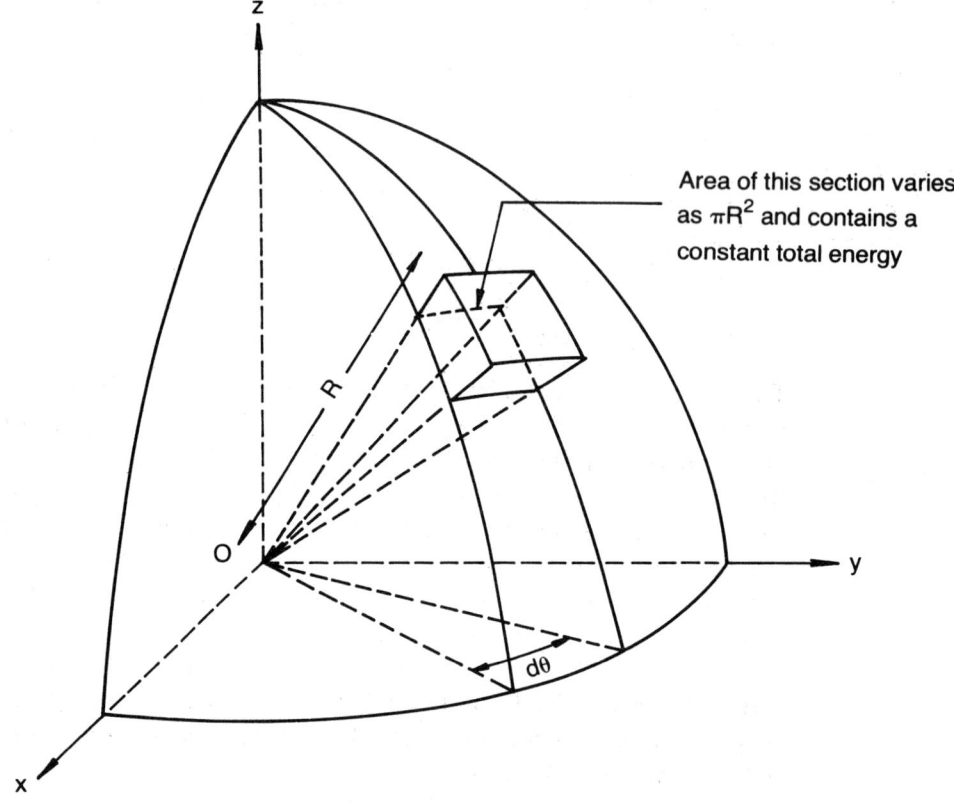

Figure 4.1   *The originating energy from the origin, "O," is dispersed uniformly over a spherical surface as it propagates in space.*

be shown later in this chapter that, although the attenuation is an increasing function of distance, it is not a very strongly frequency-dependent function.

However, because the capture area, or aperture, of a mobile antenna decreases directly with frequency, the energy captured by the antenna is directly a function of frequency. Thus, a quarter-wave antenna at 450 MHz is twice as long as a 900-MHz antenna and so can capture more energy from a field with the same intensity. This difference in capture area or aperture is what mainly accounts for the better long-range performance of lower-frequency systems.

Of course, this discussion must be limited to a frequency band where the propagation mode is similar. If the range 150–1000 MHz is considered, then the assumptions will hold.

Mathematically, the effective aperture, or capture area A, of an antenna is

$$A_{eff} = \frac{\lambda^2 G}{4\pi}$$

where

λ = wavelength

G = antenna gain

A_{eff} = effective aperture

Therefore, if the wavelength is decreased by a factor of 2, the effective capture area will increase by a factor of 4. The energy received will also increase by a factor of 4. Thus, the power collected by a 450-MHz dipole antenna compared to a 900-MHz dipole antenna is 10 log 4 = 6 dB higher in a field of the same intensity. The same result can be derived by visualizing the antenna as immersed in an electric field of V volts per meter. A longer antenna is swept by more lines of field strength and so induces a higher voltage.

If a 900-MHz antenna is compared to a 450-MHz antenna of the same type, the 450-MHz antenna is twice as long and will intercept twice the electric field potential. The resulting increase in received field strength is 20 log 2 or 6 dB.

The effective aperture of an antenna of the same gain (for example, 3 dB) will also slightly effect the relative performances of 800-MHz and 900-MHz systems.

Field strength is a measure of power density at any given point. The main units of field strength are dBμV/m, dBμV, μV, and dBμ. The dBμV/m unit measures power density of radio waves. The other units measure received energy levels. Most mobile-radio engineers prefer the dBμV/m unit, while microwave engineers prefer dBm, and bench technicians use μV. This preference is partly historical and partly the practicality of the units in the different fields.

To visualize how these units relate, consider Figure 4.2, which shows a test dipole in the plane at right angles to the direction of propagation. It measures the electric field strength in the direction of the antenna.

The units dBμV, dBm, and μV measure the power or voltage received by a dipole antenna in the field. Voltages in mobile equipment are usually measured at 50 ohms, but any impedance can be used. These units are defined as

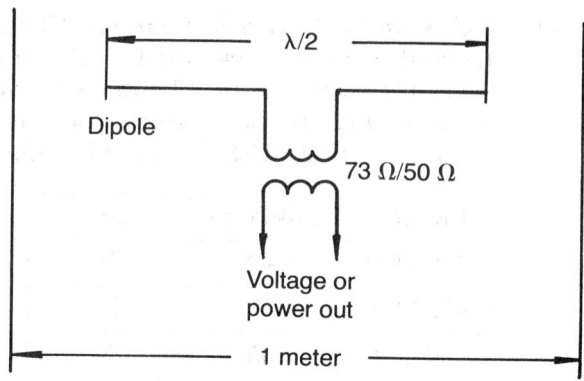

Figure 4.2   *The concept of measurement of field strength in μV/m using a dipole.*

dBμV =

$$20 \log \frac{\text{voltage at transformer output left (terminated in 50 ohms)}}{1 \text{ microvolt}}$$

$$\text{dBm} = 10 \log \frac{\text{power transformer}}{1 \text{ milliwatt}} \text{ (impedance independent)}$$

μV = voltage at transformer terminal into 50 ohms load in microvolts

The unit dBμV/m is the voltage potential difference over 1 meter of space, measured in the plane at right angles to the direction of propagation and in the direction of the test antenna. It is defined as follows:

$$\text{dBμV/m} = 20 \log \frac{\text{voltage potential 1 meter}}{1 \text{ microvolt}}$$

To see how all these units relate to each other in a real environment, a case study based on the work of Okumura et al. (a classic work on mobile propagation) will help.

Consider a typical base station that has a reference height of 200 meters and a transmitter ERP of 100 watts. Table 4.1 lists the received field strength for various transmitter frequencies in the far field; for example, 10 km.

From the table, it should be clear that the actual receiver signal varies enormously with frequency, even though the transmitter power, site, and antenna height remain fixed. Therefore, if the units μV, dBm, or dBμV are selected, the results of a survey of one fre-

Table 4.1   *Field strength at 10 km for a 100-watt TX at 20 meters in an urban environment. (Note: This table was derived from a paper by Okumura et al. entitled "Field Strength and Its Variability in VHF and UHF Land-Mobile Radio Service," Review of the Electrical Communication Laboratory, Vol. 16, Nos. 9, 10, Sept-Oct 1968, pp. 835–873.)*

FREQ	dBμV/m	μV	dBm	dBμV
150 MHz	49	71	–70	+37
450 MHz	47	17	–82	+25
900 MHz	45	8	–89	+18

quency cannot easily be translated to frequencies that are significantly different. However, if dBμV/m is selected, only a few dB separate the readings. Furthermore, if additional sites are studied (that is, different heights for the transmitter, and different distances), this relationship is retained. Thus, any field strength measured in dBμV/m measures energy density at a given point and is dependent on frequency, only to the extent that atmospheric and clutter attenuation is dependent on frequency. Therefore, it is possible to use results from a survey done at one frequency to draw conclusions about another if an allowance of 2 dB per octave is used. Some caution should be exercised when using these broad generalizations. However, they can be most useful approximations of coverage.

RELATIONSHIP BETWEEN UNITS OF FIELD STRENGTH AT THE ANTENNA TERMINALS

Assuming a 50-ohm termination, a dipole receiving antenna (unity gain), and a zero-loss feeder, the relationship between units of field strength at the antenna terminals is as follows:

$$E(\mu V/m) = \frac{\mu V}{39.3924} \times FREQ \text{ (MHz)}$$

Equation 4.1. Starting with dBm all into 50 Ω

$$\mu V = 2.236 \times 10^5 \times 10^{\text{ dBm}/20}$$

$$dB\mu V/m = 20 \log \left(5.676 \times 10^3 \times FREQ \text{ (MHz)} \times 10^{\text{ dBm}/20}\right)$$

$$dB\mu V = 20 \log \left(2.236 \times 10^5 \times 10^{\text{ dBm}/20}\right)$$

Equation 4.2. Starting with μV all into 50 Ω

$$dBm = 20 \log \frac{\mu V}{2.236 \times 10^5}$$

$$dB\mu V = 20 \log \mu V$$

$$dB\mu V/m = 20 \log (\mu V \times FREQ \text{ (MHz)}/39.3924)$$

Equation 4.3. Starting with dBμV/m all into 50 Ω

$$dB\mu V = dB\mu V/m - 20 \log \frac{FREQ\text{(MHz)}}{39.3924}$$

$$dBm = 20 \log \left(\frac{10^{dB\mu V/m/20} \times 1.76168 \times 10^{-4}}{FREQ \text{ (MHz)}} \right)$$

$$\mu V = \frac{39.3924 \times 10^{\ dB\mu V/m/r20}}{FREQ \text{ (MHz)}}$$

Equation 4.4. Starting with dBμV all into 50 Ω

$$\mu V = 10^{dB\mu V/20}$$

$$dBm = 20 \log \frac{10^{dB\mu V/20}}{2.236 \times 10^5}$$

$$dB\mu V/m = dB\mu V + 20 \log \frac{FREQ \text{ (MHz)}}{39.3924}$$

CONVERSION TABLES

Because it is very easy to make a mistake when applying the formulas to translate between units, the conversion table in Table 4.2 can be very helpful. This table is useful for calculating the relationship between the variables illustrated in Figure 4.2. If a different antenna impedance is considered (a 300-ohm folded dipole antenna, for example), then the results cannot be used directly. Similarly, in a real-life environment, the signal levels will probably be measured at the receiver output, as shown in Figure 4.3, and the necessary corrections must be applied.

The relationship between the variables must be adjusted by the antenna gain minus the cable loss. At 900 MHz, using a 3-dB antenna, the cable loss if 3 meters long is about 3 dB and thus the correction factor approaches zero. At other frequencies, this approximate relationship will not hold.

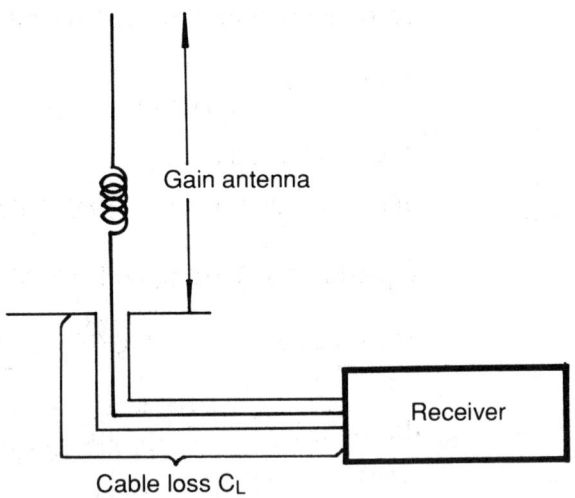

Figure 4.3 *Actual field strength measurement.*

Table 4.2 *Conversion tables for a 50 Ω dipole antenna (for 150 MHz, 450 MHz, and 900 MHz) (continued next page)*

			FREQ. MHz (dBμV/m)		
dBm	**μV**	**dBμV**	**150**	**450**	**900**
10	707107	117	129	138	144
1	250891	108	120	129	135
0	223607	107	119	128	134
–1	199290	106	118	127	133
–2	177617	105	117	126	132
–3	158301	104	116	125	131
–4	141080	103	115	124	130
–5	125743	102	114	123	129
–6	112069	101	113	122	128
–7	99881	100	112	121	127
–8	89019	99	111	120	126
–9	79338	98	110	119	125
–10	70710	97	109	118	124

Table 4.2   *Conversion tables for a 50 Ω dipole antenna (for 150 MHz, 450 MHz, and 900 MHz) (continued next page)*

dBm	μV	dBμV	FREQ. MHz (dBμV/m)		
			150	**450**	**900**
−11	63021	96	108	117	123
−12	56167	95	107	116	122
−13	50059	94	106	115	121
−14	44615	93	105	114	120
−15	39763	92	104	113	119
−16	35439	91	103	112	118
−17	31585	90	102	111	117
−18	28150	89	101	110	116
−19	25089	88	100	109	115
−20	22360	87	99	108	114
−21	19929	86	98	107	113
−22	17761	85	97	106	112
−23	15830	84	96	105	111
−24	14108	83	95	104	110
−25	12574	82	94	103	109
−26	11206	81	93	102	108
−27	9988	80	92	101	107
−28	8902	79	91	100	106
−29	7933	78	90	99	105
−30	7071	77	89	98	104
−31	6302	76	88	97	103
−32	5616	75	87	96	102
−33	5005	74	86	95	101
−34	4461	73	85	94	100
−35	3976	72	84	93	99
−36	3543	71	83	92	98
−37	3158	70	82	91	97

Table 4.2 *Conversion tables for a 50 Ω dipole antenna (for 150 MHz, 450 MHz, and 900 MHz) (continued next page)*

			FREQ. MHz (dBµV/m)		
dBm	µV	dBµV	150	450	900
−38	2815	69	81	90	96
−39	2508	68	80	89	95
−40	2236	67	79	88	94
−41	1993	66	78	87	93
−42	1776	65	77	86	92
−43	1583	64	76	85	91
−44	1410	63	75	84	90
−45	1257	62	74	83	89
−46	1120	61	73	82	88
−47	998	60	72	81	87
−48	890	59	71	80	86
−49	793	58	70	79	85
−50	707	57	69	78	84
−51	630	56	68	77	83
−52	562	55	67	76	82
−53	501	54	66	75	81
−54	446	53	65	74	80
−55	398	52	64	73	79
−56	354	51	63	72	78
−57	316	50	62	71	77
−58	281	49	61	70	76
−59	251	48	60	69	75
−60	223	47	59	68	74
−61	199	46	58	67	73
−62	177	45	57	66	72
−63	156	44	56	65	71
−64	141	43	55	64	70

Table 4.2   *Conversion tables for a 50 Ω dipole antenna (for 150 MHz, 450 MHz, and 900 MHz) (continued next page)*

			FREQ. MHz (dBμV/m)		
dBm	μV	dBμV	150	450	900
−65	125	42	54	63	69
−66	112	41	53	62	68
−67	100	40	52	61	67
−68	89	39	51	60	66
−69	79	38	50	59	65
−70	70	37	49	58	64
−71	63	36	48	57	63
−72	56	35	47	56	62
−73	50	34	46	55	61
−74	45	33	45	54	60
−75	39	32	44	53	59
−76	36	31	43	52	58
−77	32	30	42	51	57
−78	28	29	41	50	56
−79	25	28	40	49	55
−80	22	27	39	48	54
−81	20	26	38	47	53
−82	18	25	37	46	52
−83	16	24	36	45	51
−84	14	23	35	44	50
−85	13	22	34	43	49
−86	11	21	33	42	48
−87	10	20	32	41	47
−88	9	19	31	40	46
−89	8	18	30	39	45
−90	7	17	29	38	44
−91	6.3	16	28	37	43

Table 4.2 *Conversion tables for a 50 Ω dipole antenna (for 150 MHz, 450 MHz, and 900 MHz) (continued next page)*

dBm	μV	dBμV	FREQ. MHz (dBμV/m)		
			150	**450**	**900**
−92	5.6	15	27	36	42
−93	5	14	26	35	41
−94	4.5	13	25	34	40
−95	4	12	24	33	39
−96	3.5	11	23	32	38
−97	3.2	10	22	31	37
−98	2.8	9	21	30	36
−99	2.5	8	20	29	35
−100	2.2	7	19	28	34
−101	2	6	18	27	33
−102	1.8	5	17	26	32
−103	1.6	4	16	25	31
−104	1.4	3	15	24	30
−105	1.3	2	14	23	29
−106	1.1	1	13	22	28
−107	1	0	12	21	27
−108	0.89	−1	11	20	26
−109	0.79	−2	10	19	25
−110	0.71	−3	9	18	24
−111	0.63	−4	8	17	23
−112	0.56	−5	7	16	22
−113	0.5	−6	6	15	21
−114	0.47	−7	5	14	20
−115	0.4	−8	4	13	19
−116	0.35	−9	3	12	18
−117	0.32	−10	2	11	17
−118	0.28	−11	1	10	16

Table 4.2   *Conversion tables for a 50 Ω dipole antenna (for 150 MHz, 450 MHz, and 900 MHz) (continued)*

			FREQ. MHz (dBμV/m)		
dBm	**μV**	**dBμV**	**150**	**450**	**900**
−119	0.25	−12	0	9	15
−120	0.22	−13	−1	8	14
−121	0.2	−14	−2	7	13
−122	0.17	−15	−3	6	12
−123	0.16	−16	−4	5	11
−124	0.14	−17	−5	4	10
−125	0.126	−18	−6	3	9
−126	0.122	−19	−7	2	8
−127	0.1	−20	−8	1	7
−128	0.089	−21	−9	0	6
−129	0.079	−22	−10	−1	5
−130	0.071	−23	−11	−2	4

STATISTICAL MEASUREMENTS OF FIELD STRENGTH

In point-to-point radio, field strength is a one-dimensional variable of time. Most of the time variance is due to log-normal fading, and the nature of the signal variability is well documented. Because it is a simple function of time, field strength in point-to-point radio can be easily understood and measured. The situation is somewhat more complex in the mobile RF environment. Thus the field strength in the point-to-point environment is a simple function of time as shown here:

$$F = f(t)$$

The situation is somewhat more complex in the mobile environment where the field strength also varies with location (space), and so the measured value is a four-dimensional statistical variable

$$F = f(t,x,y,z)$$

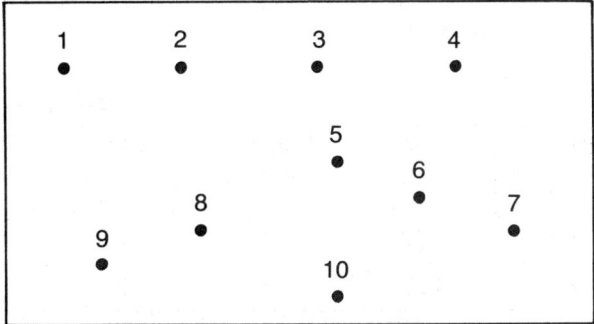

Figure 4.4 *The measurement of field strength over an area is made at unique points in space and time. Actual readings vary with time (due to log-normal fading) and space (due to log-normal and Rayleigh fading). Any statement about the field strength in this area is a statement about a variable of space and time.*

or, if $(x, y, z) = L$ = location, then

$$F = f(t, L)$$

The real-life measurement of the field strength of an area is the collective result of a number of measurements made in different points in space and time, as illustrated in Figure 4.4. Thus, if a measure of field strength is required to typify an area, then its statistical nature dictates that the actual result depends on when and where the measurement is made. Let's assume a measurement of the average field strength along a 500-meter section of a road is required. If all samples are taken at the Nyquist rate (in space at locations $L_0, L_1, \ldots L_m$) and at one instant (T_0), the result would be an average value at a particular time, T_0.

Mathematically this can be expressed as

$$F_0 = \frac{\displaystyle\sum_{i=0}^{m} f(T_0, L_i)}{m + 1}$$

($m + 1$ = the number of readings taken at positions $L_0, L_1, \ldots L_m$)

In practice making simultaneous measurements is difficult and the individual space measurements would be made at different times ($T_1 \ldots Tn$). The average value in time and space of the field strength in that region is then given by F as shown here:

$$\frac{\sum_{K=0}^{n} F_K}{n}$$

Notice that since Nyquist-rate space samples were taken, it is not necessary to use different locations for samples in space; but this could be done. Virtually all real measurements are an average (or some other statistical measure) of samples in different locations in space and time.

Other statistical measures are widely used. For example, all readings can be recorded and separated into two equal groups depending on level. Thus, it is possible to obtain the value above (or below) which 50 percent of the recorded samples occur. This is known as the median value (not the same as the average value). If all readings were taken at a single instant, the median value F_{50} can be found and the result would be at time T_0

$$F_{50} = f \text{ median } \{f(T_0, L_i)\} \text{ for } i = 0 \text{ to } m$$

where

f_{50} = the level at which 50 percent of the samples 0 to m lie below

In practical terms, as survey readings are taken at different points in space and time for a given region, the average of a set of readings in a region is the true time/space averaged measurement. Because time fluctuations of field strength are usually fairly fast relative to measurement periods, this assumption holds fairly well. Lognormal fades produce field-strength variations with a periodicity of a few seconds, while a measurement is usually taken over 1- to 5-minute intervals. Therefore, if measurements taken from a moving vehicle in one survey record F_{50} field strength, it is considered to be the field strength for 50 percent of locations and 50 percent of the time.

As the number of samples in space equals the number of samples in time, it is also reasonable to use the same technique to obtain the 70 percent/70 percent or 90 percent/90 percent values. However, because F_{50} is a function of space and a different one of time, a simple survey technique cannot be used to directly obtain the value for 90 percent of locations for 50 percent of the time. To obtain this value it would be necessary to obtain the standard deviation of the time-dependent variation independently (approximately the standard deviation of the log-normal fade).

Interpreting a paging requirement for the field strength to be, for example, 32 dBµV/m for 90 percent of locations and 90 percent of the time is a difficult task because the concept has never been adequately defined. In particular, the concept originally applied to small regions of space and its application to service areas is vague. It might mean that within the service area only 10 percent of all readings will be below 32 dBµV/m and this is how most people think of it. However, the goal of a system designer is not to have 10 percent of the service area substandard. Probably what is really meant by this is that 10-percent substandard coverage is a worst-case scenario and is an upper limit rather than a design parameter. One would not expect, for example, a batch of resistors with 10-percent tolerance to be *designed* to be 10 percent out of tolerance. The 10-percent limit merely means the manufacturer's tolerance or worst case is 10 percent.

 # RADIO SURVEY

A radio survey is the process of measuring the propagated radio field strength over an area of interest. It is an essential part of the paging site-selection process. Many radio-survey techniques exist, but few yield consistent and satisfactory results. A radio survey is necessary as a design aid and as a maintenance tool. As a design aid, it helps determine potential coverage of a proposed base station site. As a maintenance tool, a radio survey confirms continued satisfactory coverage.

A radio survey usually uses a field-strength measuring receiver located in a vehicle to measure the field strength. Sometimes the reciprocal path (that is, the path from the mobile to the base station) is measured instead. Both measurements are mathematically equivalent.

When measuring field strength, it is important to note that what is being measured is a statistical variable and that the measurement technique must allow for this.

Three factors operate together to produce the measured field strength: path loss (free space), log normal fading, and Rayleigh fading. Figure 5.1 illustrates the free (for example, unobstructed) path loss. This loss is the most significant in microwave links, but it is only one of the losses in the mobile environment.

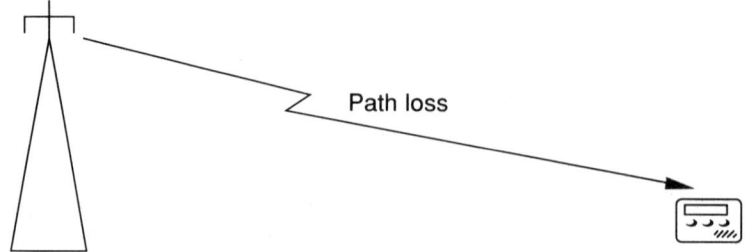

Figure 5.1   *Path loss (free space).*

The free space path loss P_L is given by:

$$P_L = 20 \log (42 \cdot d_{km} \cdot f_{MHz}) \text{ dB} = 32.5 + 20 \log f_{MHz} + 20 \log d_{km} \text{ dB}$$

where

P_L = path loss in dB

d = distance

d_{km} = distance in kilometers

f_{MHz} = frequency in megahertz

Figure 5.2 shows log normal fading. This process is called log normal fading because the field-strength distribution follows a curve that is a normally distributed curve, provided the field strength is measured logarithmically.

Multipath, or Rayleigh, fading is a salient feature of mobile communications and, to some significant extent, limits the coverage of mobile systems when the mobile is moving in a multipath environment. It is not such a dominant factor in handheld mobile usage but, in low-field-strength areas, it can be detected by variations in noise levels as the receiver is moved. Figure 5.3 illustrates multipath fading.

Figure 5.2   *Log normal fading that is due to obstruction is known as "shadowing" or "diffraction losses."*

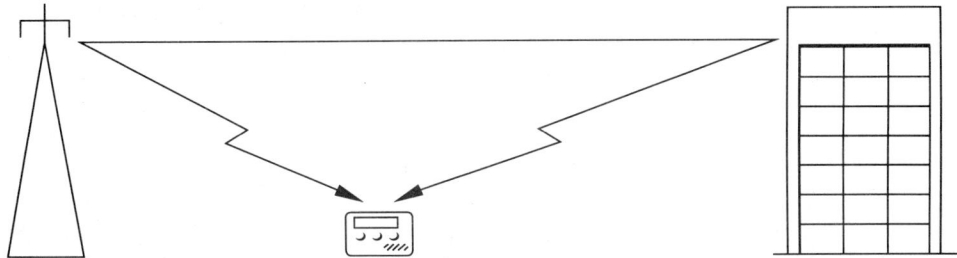

Figure 5.3 *Multipath, or Rayleigh, fading is produced as a result of interference patterns between the various signal paths in a multipath environment.*

An empirical formula for the cumulative effect of these three types of fading is given in *Recommendations and Reports of the CCIR, 1982, Volume V, Report 567-2* as

$$P_L = 69.55 + 26.16 \log f_{MHz} - 13.82 \log h_1 - a\,(h_2)$$

$$+ (44.9 - 6.55 \log h_1) \log d_{km} \text{ dB}$$

where

P_L = loss in dB

f_{MHz} = frequency in megahertz

h_1 = base station antenna height in meters

h_2 = receiver antenna height in meters

$a(h_2) = (1.1 \log f - 0.7)\, h_2 - (1.56 \log f - 0.8)$

d_{km} = distance in kilometers

where 15 percent of the area is covered by buildings (that is, an urban area).

This formula is based on field experience. Experience dictates that in different terrains some or all of the coefficients must be recalibrated. For general use, the formula should be regarded as accurate to ±10 dB.

It is often said that in the mobile environment, the loss is inversely proportional to distance to the fourth power. You can see that this is consistent with the formula by looking at the last term (44.9 − 6.55 log h_1) log d_{km}. This is the only term that is a function of d_{km}. If, for example, h_1 = 30 meters (the base-station antenna height), then this term becomes (44.9 − 6.55 log 30) × log d_{km} = 35.2 log d_{km}.

Because this is an expression for loss in dB, it can be rewritten in the form

$$\text{Loss} \, \alpha \, \frac{1}{d_{km}^{3.52}}$$

This reduces the relatively complex formula to approximately d_{km}^{4}; the fourth power relationship holds exactly at an antenna elevation of 5.6 meters.

STANDING WAVE PATTERNS

In the far field, where all these loss modes are operating, the field strength varies with distance and time. At any one instance, the field strength can be shown as illustrated in Figure 5.4.

The limiting case of standing wave patterns is one produced by a reflecting plane at right angles to the line of propagation. A standing wave produced by a wave incident on a plane reflecting surface (such as a wall) produces the familiar $\lambda/2$ standing wave pattern shown in Figure 5.5.

Other forms of interference generally produce interference patterns with a wavelength greater than $\lambda/2$. The distance L, between the waves, is such that $\lambda/2 < L$, but L can take any larger value. In practice, however, $L \approx \lambda/2$ can be taken as the worst case. Notice that $\lambda/2$ = 0.16 m (16 cm) at 900 MHz.

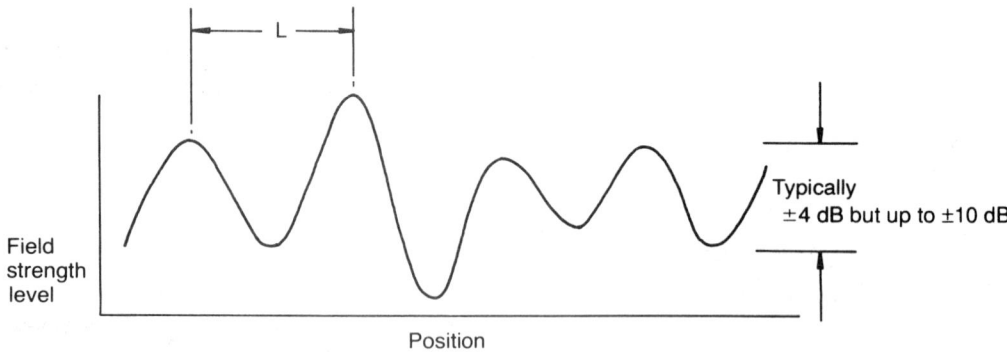

Figure 5.4   *Standing wave pattern.*

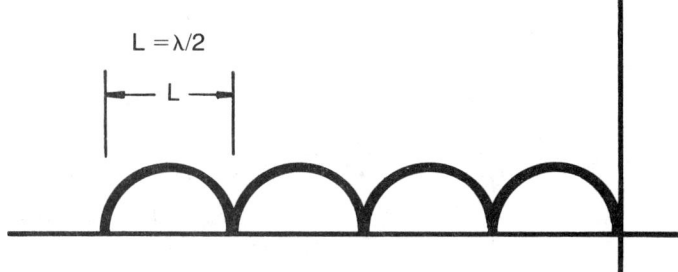

Figure 5.5 *Standing wave pattern caused by reflections off a plane surface.*

MEASURING FIELD STRENGTH

As a mobile radio must necessarily operate over its entire service area, the field strength at a point becomes meaningless in terms of the overall performance of a mobile receiver. Consequently, individual spot readings are also meaningless.

Some operators have tried to solve this problem by using a field-strength meter in a moving vehicle and "guessing" the average level. This also yields meaningless results, as you can see from the structure of a typical field-strength meter, as shown in Figure 5.6.

The field-strength meter consists of a receiver, which has some way to access the IF limiter or AGC (Automatic Gain Control) drive stage to measure the output of that stage via a meter. A log-law amplifier gives a usable output in dB. The meter is usually a conventional moving-coil meter.

Because these field-strength meters are designed to operate in a point-to-point environment, the smoothing capacitor C (or its

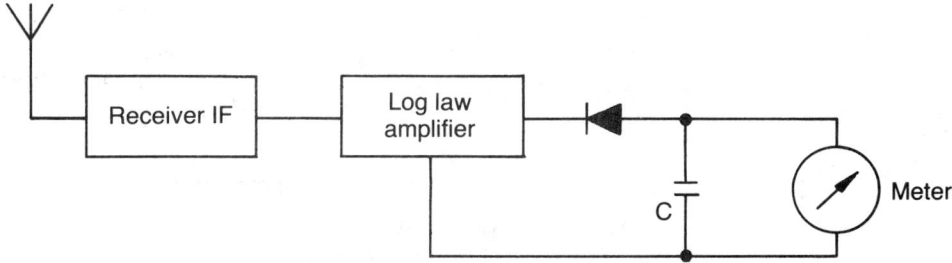

Figure 5.6 *Basic field-strength meter.*

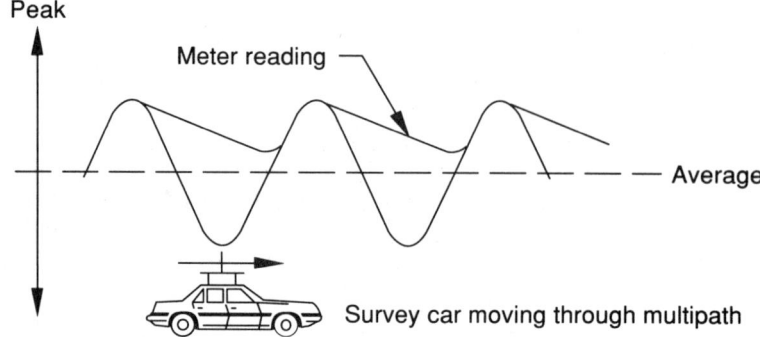

Figure 5.7   *Detector response.*

mechanical equivalent) is usually incorporated to even out small fluctuations due to log normal fading, and it likely has a time constant of 0.5–2 seconds.

When such a meter is confronted with a rapidly varying signal, it tends to follow the peaks, as indicated in Figure 5.7.

Because the relationship between the peak value and the average value from such a device is dependent on the depth of the fade and the multipath frequency, no firm relationship between these values can be said to exist. Thus, a sampling method that samples at a sufficiently high rate to measure the actual standing wave is necessary.

If this standing wave pattern were totally uncorrelated, the necessary sampling speed would be the Nyquist rate (2 times the pattern frequency), but, because of the existence of a correlation, in practice, it has been demonstrated that about one quarter of that rate yields results that are accurate to ±1 dB.

SAMPLING SPEED

To calculate the required sampling speed in a mobile environment, consider a vehicle moving at 100 km per hour through a standing wave pattern of a 900 MHz transmission.

The wavelength of that pattern is

$$C/F = \frac{300,000,000}{900,000,000} = 0.333 \text{ meters}$$

where

C = speed of light

F = frequency

In the worst instance, the standing wave pattern is repeated every $\lambda/2$ m (as in the case of a reflection of 180 degrees from a wall).

So $\lambda/2 = 0.333/2 = 0.1665$ meters

100 km per hour = 27.7 meters/sec.

The Nyquist sampling rate requires two samples per pattern interval, or one sample every 0.1665/2 meters. Thus, the Nyquist sampling rate is $27.7/(0.1665/2) = 332$ samples/sec. As already mentioned, about one quarter of that rate (or 80 samples/sec) would suffice.

Quite a number of commercially available field strength meters sample much slower than 80 samples/sec, and will consequently return inaccurate readings if used in a moving vehicle. Notice that the inaccuracy will increase as the sample speed decreases and that, by its nature, the error will be randomly distributed. At lower frequencies proportionally lower sampling rates are permissible.

MODERN SURVEY TECHNIQUES

Modern survey equipment is based on the principles of the system illustrated in Figure 5.8. The radio receiver can be a specialized communications receiver with a wide dynamic range of both level and frequency, or it can be a special-purpose receiver such as a mobile telephone. If the receiver is not designed as a measuring receiver, some limitations will occur due to the limiter time constants.

The two main criteria for the receiver are that the RF level is a monotonic (single-valued) function of the RF level and that the limiter output has a bandpass characteristic of about 100 Hz.

Most modern receivers meet the requirement that the RF level is monotonic. Some older wide-band communications receivers with a measuring capability use switched attenuators to increase their dynamic range. Often the switching-in of these attenuators causes discontinuities in the output level, which can render them unsuitable for survey.

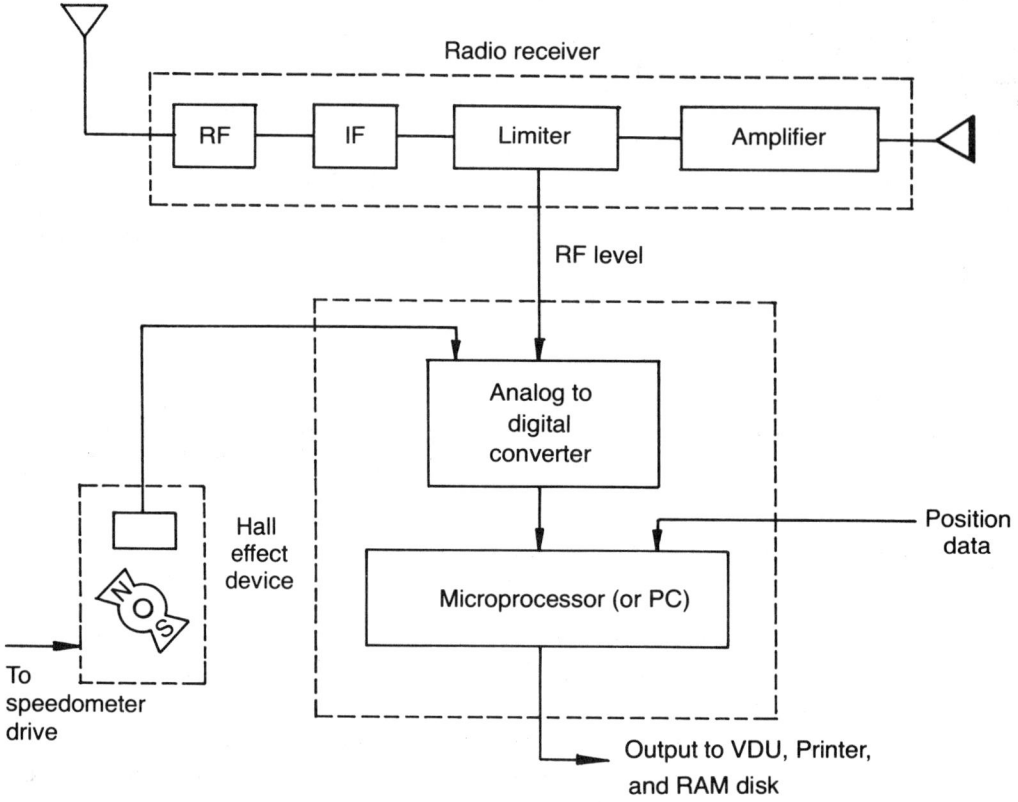

Figure 5.8   *Basic survey block diagram.*

The response time of the limiter is deliberately damped so that the device operates only to compensate for level variations that ordinarily result from fading, but the response frequency is limited so that RTTY, MORSE, and other digital signals can be passed normally. This damping can be as elementary as a simple RC bandpass filter, or it may be more sophisticated.

Figure 5.9 shows the limiter/AGC response curve. If it was originally planned that the receiver was to operate in a fixed location, this damping may limit the response time of the limiter to the extent that it is unsuitable for survey.

The easiest way to determine suitability is to input an AM square-wave modulated carrier and compare the input with the limiter drive as the frequency is increased. Provided the output tracks fairly well up to 50 Hz, the receiver will be effective. Figure 5.10 shows the limiter response time of a survey receiver.

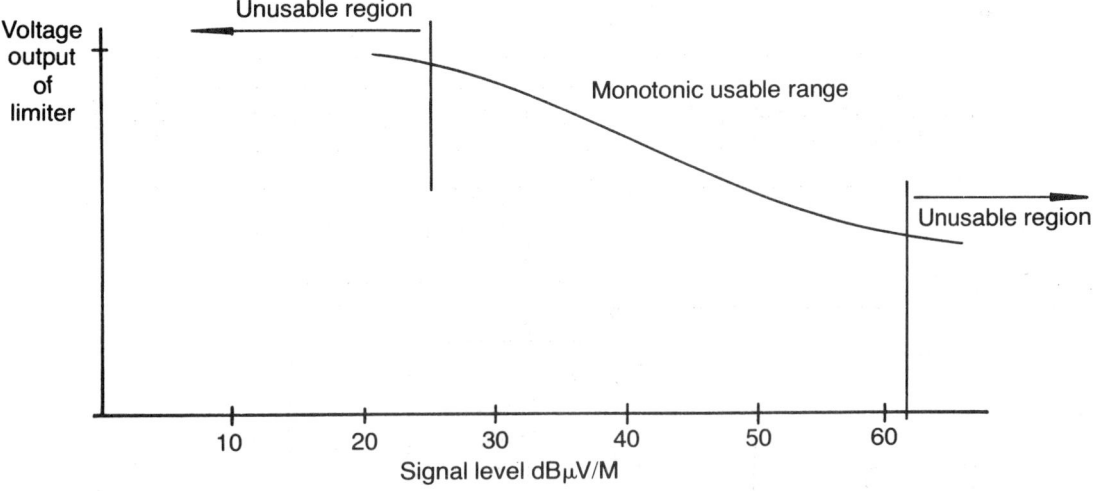

Figure 5.9   *Usable portion of limiter/AGC response curve.*

Most measuring receivers track only up to 50–100 Hz, although a few go a little higher than this. Notice that the inability to sample to at least 50 Hz results in damping errors of the same nature as those of a simple field-strength meter previously discussed.

In a digital field-strength meter, the output of the limiter is read by an analog-to-digital converter, which samples the limiter levels. As previously explained, the sample rate depends on the RF frequency and vehicle velocity. Figure 5.11 shows a typical A/D converter.

A single high speed analog-to-digital converter chip measures the limiter output voltage and outputs the level to a data bus. Usually, a multiplexer chip is used to allow multiple inlets to be sampled in turn.

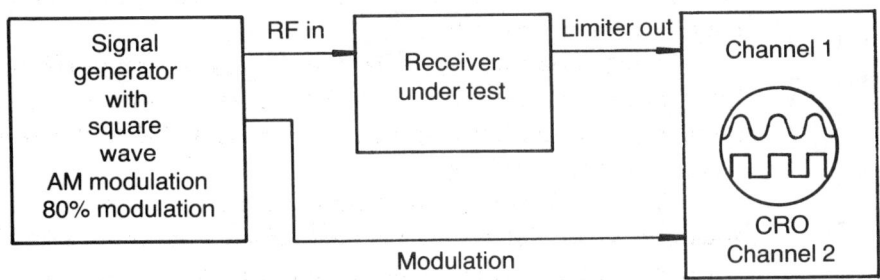

Figure 5.10   *Testing the limiter response time of a survey receiver.*

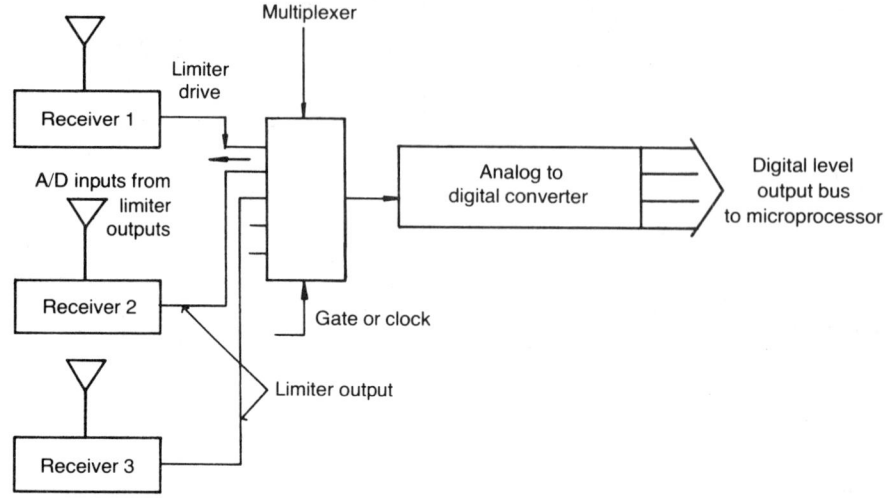

Figure 5.11   *An A/D converter with multiple inputs from a bank receiver.*

Receivers not specifically designed as measuring receivers may well have a band pass characteristic similar to the one shown in Figure 5.12. You can see that the limiter drive does not respond to low-frequency signal variations. Such a receiver will probably "see" Rayleigh fading but will not respond to log normal fades. Such a receiver may also have a square-wave response, as illustrated in Figure 5.13. Notice that the behavior of the receiver in Figure 5.13 seems to be limited mainly by a simple RC network.

These receivers also have a response to large-level changes that saturate very readily at excursion of about ±4–6 dB. Thus, they generally underestimate the standard deviation and the decile value.

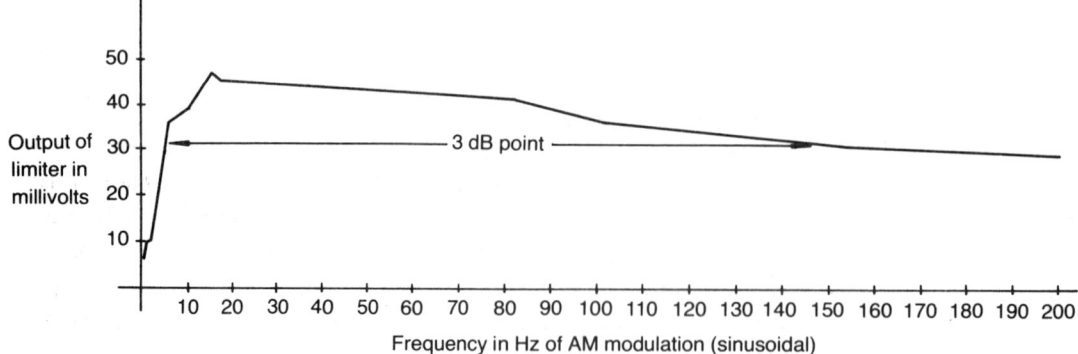

Figure 5.12   *Frequency response (expressed in Hz of AM modulation-sinusoidal) of the limiter voltage with respect to a constant carrier level change. (An ideal measuring receiver has a flat response.)*

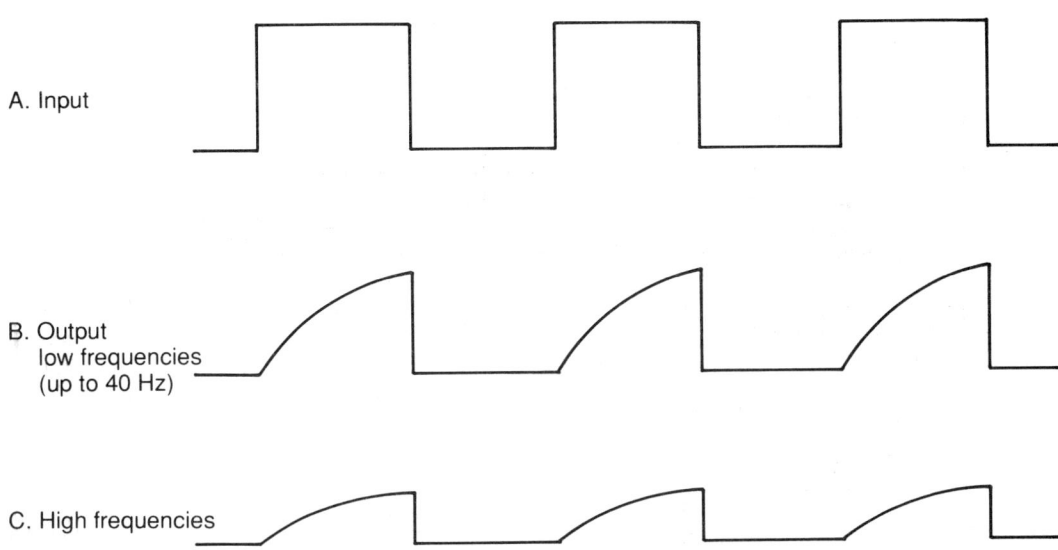

A. Input

B. Output
   low frequencies
   (up to 40 Hz)

C. High frequencies

Figure 5.13   *The limiter drive output response of a commercial wide-band receiver that was not designed to measure a square wave.*

The speedometer pulses can be used to ensure that samples are taken regularly over fixed-distance intervals. This is done to avoid the sample biases that can occur when, for example, the survey vehicle is held up at one spot (a traffic light, for example) and so records many readings at that one point, which then give a distorted average value for the area.

The speedometers in many modern vehicles are driven by a Hall effect device (see Figure 5.8) that has a sufficiently high output level to be read directly by the A/D card by using it to gate the A/D card.

Position data can also be read and recorded along with the readings.

SAMPLING INTERVAL

Within localized regions, the average field strength as measured from a distant transmitter can be shown to be approximately constant over intervals of 50 m to 500 m. Within these localized regions, the Rayleigh and log normal fades occur; the log normal mode generally dominates, although significant Rayleigh fading may occur, particularly in built-up areas.

Studies by Okumura *et al.* have shown that the local field strength in an area bisected by a 50–500 meter path can reasonably be

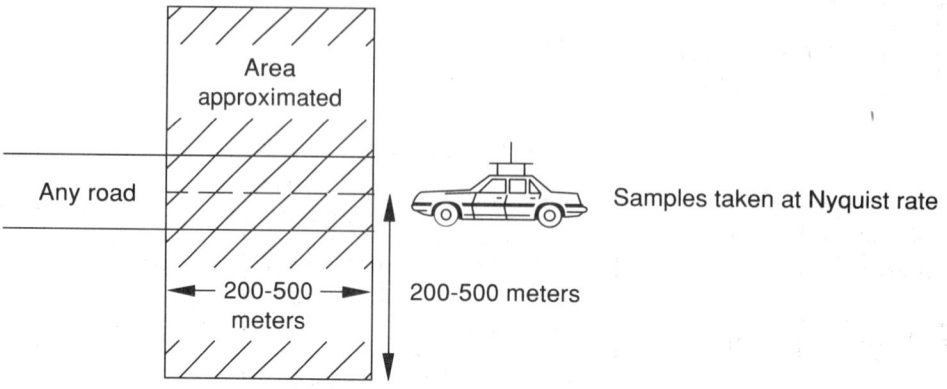

Figure 5.14   *Area represented by sample measurement.*

accurately described by a single mean (average value) and a standard deviation. Figure 5.14 shows an area represented by sample measurement.

Hence, an adequate record of regional field strength can be obtained by sampling at the Nyquist rate and obtaining the mean and standard deviation over a 50–500 meter path. This is the basic approach used by most computerized field-strength measuring devices, where the sample intervals can vary from 10 meters to 1000 meters. In practice, it can easily be shown that samples over intervals of 50–500 meters yield consistent results to a few dB.

Figure 5.15 illustrates the decile method, a commercial measuring system. The illustration shows the structure of the Radio Survey Master system from Telecom Australia, which was developed by the author. The system has an external trigger (based on the speedometer pulses) that ensures that samples are taken at fixed intervals. The system records mean, standard deviation, decile value, the number of samples taken, and local time. The decile value is often used and favored by some because it gives a more conservative estimation of mobile performance than the average value—especially in areas of deep multipath. Figure 5.16 shows a survey vehicle equipped with this system.

In Figure 5.17, the upper 10-percent decile corresponds to T; this means that 90 percent of the readings are above the level T dBμV/m.

The 90-percent level is calculated using an iteration method. If the average is also needed from this sample, you should note that, although decile levels can be determined with equal accuracy, as an absolute level or the log of the level (dBs), the values must be converted to absolute values (μV/m) before a true average can be calculated.

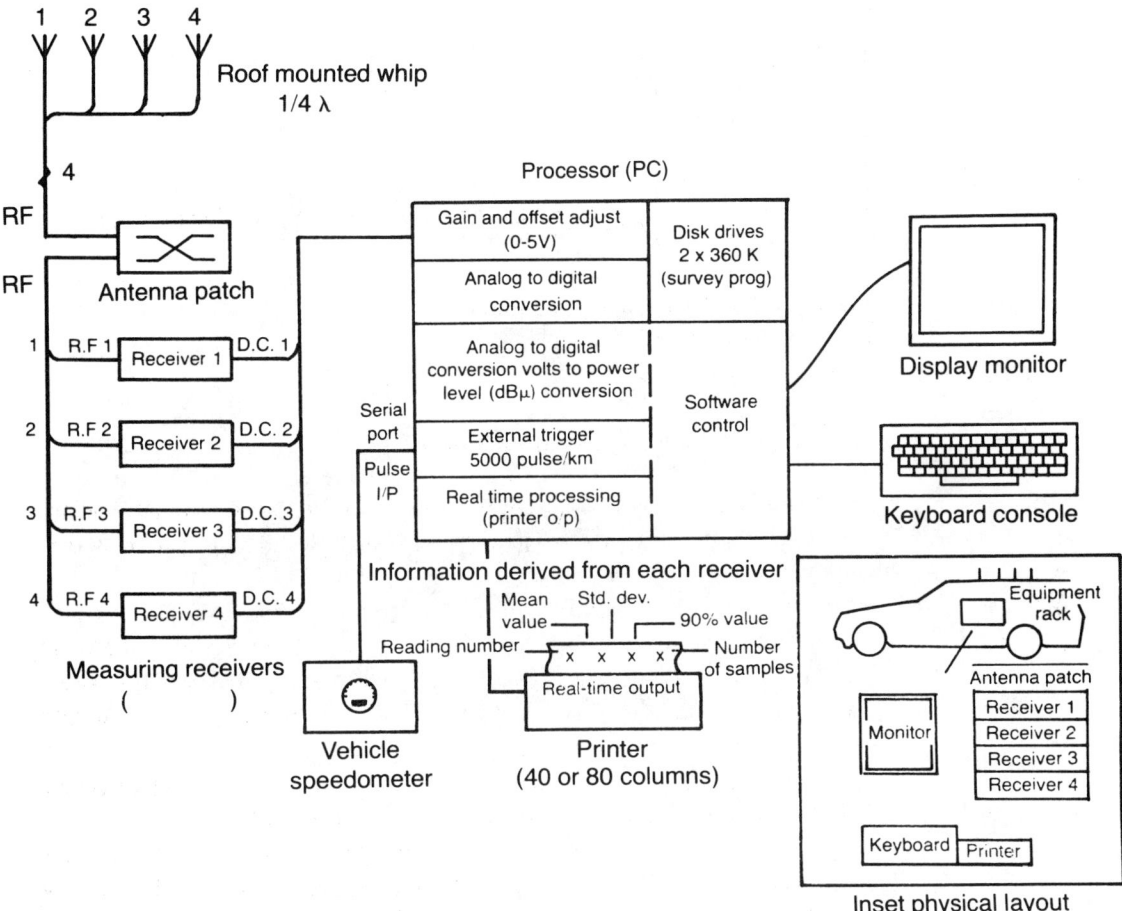

Figure 5.15 *Vehicle-mounted survey equipment.*

In high-multipath areas with deep fades (high standard deviation), a measured mean value does not tell much about the extremities of the readings and particularly about the minimum (and hence, noisy) locations. Taken with the standard deviation, more can be extracted, but the decile method eliminates the need to know about the standard deviation and allows the use of a single standard-design field strength.

Figure 5.18 shows an example of log normal distribution of field strength.

For example, if a field strength of 39 dBμV/m average with a standard deviation of 6 dB (in a suburban area) is considered the objective, this can be equated to 39 dBμV/m − 0.87 × 6 = 34 dBμV/m

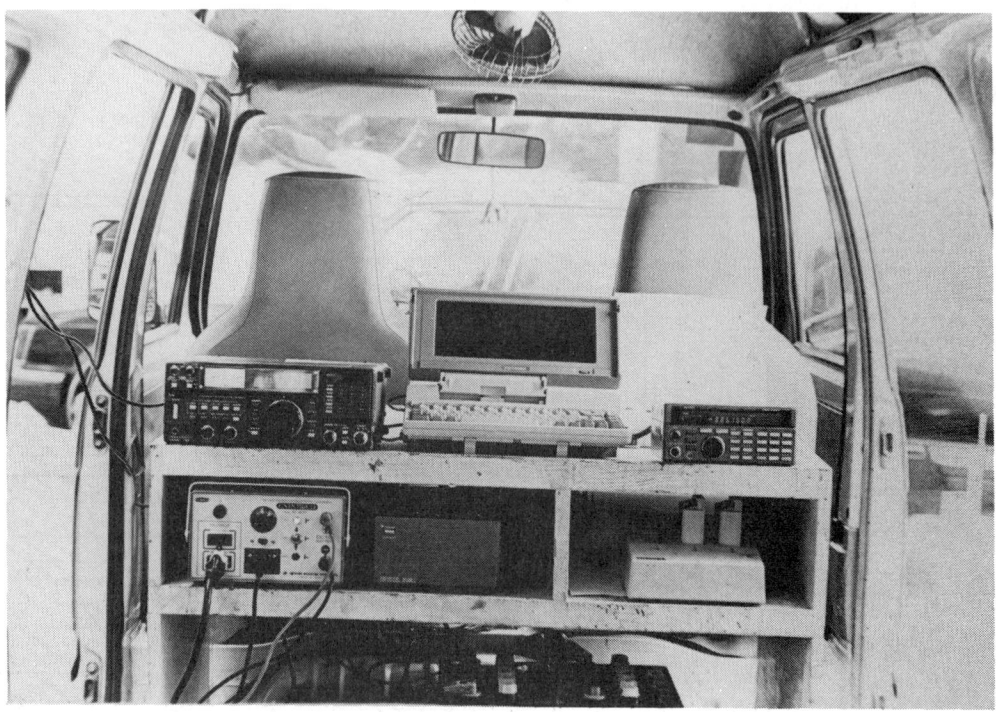

Figure 5.16   *A survey vehicle with equipment as described in Figure 5.15.*

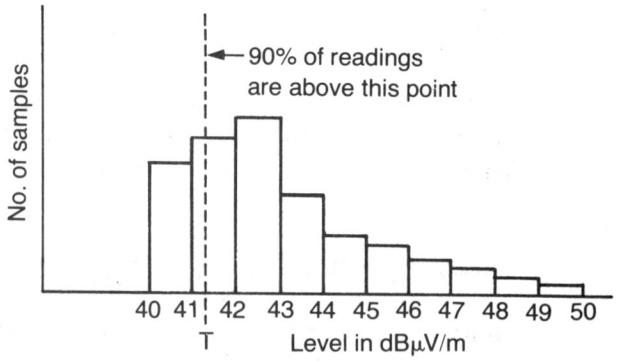

Figure 5.17   *Histogram of sampled data.*

for 90 percent of readings to define the boundary corresponding to 90 percent of locations having 39 dBμV/m average 90 percent of the time.

Because in lower multipath regions the 90-percent reading moves closer to the mean (and conversely, further away in high multi-

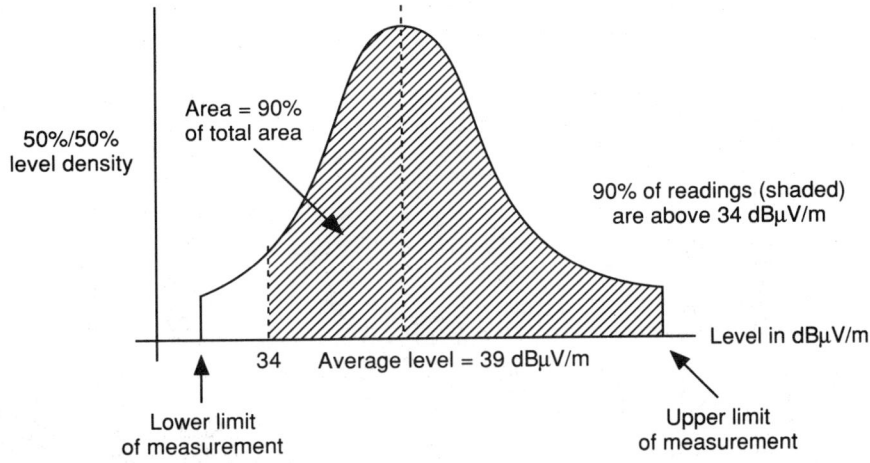

Figure 5.18 *Log normal distribution of field strength in 50 percent of location and time.*

path regions), this method adjusts to the multipath environment in a way that an average reading cannot.

For example, if the field-strength measuring equipment was set as above (34 dB for 90 percent of readings) and it were to move from a suburban area ($\sigma = 6$ dB) to a rural area where $\sigma = 2$ dB, then the measured field strength would correspond to $34 + 0.87 \times 2 = 35.7$ dBμV/m average. That is, at this lower multipath level, an adjustment is automatically made for the lower noise level.

Clearly this method gives a more effective measure of signal quality than an average reading. It is not often done, however, because it requires somewhat more "number-crunching" power than is needed to obtain an average reading.

REVERSE PATH SAMPLING

Some surveyors prefer to use the transmitter in the mobile and locate the receiving equipment in the base station. This has the advantage of less clutter in the vehicle, but has the disadvantage of requiring a very good position-location system that can relate readings to position.

A further disadvantage is that real-time outputs cannot normally be obtained. Thus, the mobile survey team does not have continuous direct contact with the receiving site and has no feedback on the survey's progress. This can lead to many problems, including time wasted surveying areas clearly out of range and time wasted surveying when equipment may be faulty.

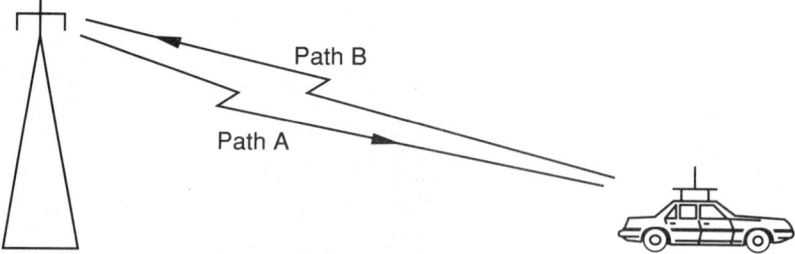

Figure 5.19   *Because the signal traversing a path to the mobile uses the same path (almost) as the signal on the path from the mobile, any losses are equal. Measuring either path yields equivalent results.*

When the measuring equipment is in the vehicle, the operators have a real-time measurement from which to solve these problems. Either way of measuring, however, if done effectively, gives equivalent results. Figure 5.19 shows that the paths *are* reciprocal and that it doesn't matter which way they are measured.

USING WIDE-BAND MEASURING RECEIVERS

Wide-band measuring receivers usually have poor sensitivity (typically 1–2 µV for 12 dB SINAD). Because surveys are often done on test transmitters with rather low ERPs (compared to the commercial paging base station), measurements are often limited by receiver performance.

It is often useful to acquire a low-noise (NF < 1dB) amplifier for the band being surveyed. Because of intermodulation susceptibility, it is best to only use such an amplifier at low signal levels and to switch it out when it is not needed. If the survey vehicle is equipped with a cellular phone (or other transmitting device), observe the effect on the performance of the measuring system; the transmitting device may desensitize the wide-band amplifier.

MULTIPLE RECEIVER ANTENNAS

Because of cross-coupling, if multiple-survey-receiver antennas are used, they should be spaced at least two wavelengths apart. Because most cellular sites are sectored, multiple receivers will effectively

Figure 5.20   *A survey vehicle can be fitted with a metal ground plane attached to ski bars. A number of antennas can be mounted on this good quality ground plane.*

increase survey efficiency. When surveying omni sites, use a second channel as a check on the integrity of the first channel.

The best way to mount multiple antennas is on a ground plane made by suspending a sheet of aluminum between the bars of a roof-rack, as shown in Figure 5.20. This not only gives a very good ground plane, but it also avoids the necessity of drilling holes in the roof of the vehicle. The whole assembly is easily detachable when the time comes to replace the vehicle.

SURVEY TRANSMITTERS

It is preferable to have the survey transmitter on the same frequency and with the same ERP as the paging base transmitter. However, this is frequently not practical, particularly because transmitters of high frequency and high-power outputs on 800+ MHz are rather inefficient and are also somewhat difficult to come by. However, at least one company, PLEXSYS, makes a dedicated survey transmitter of 45 watts

Figure 5.21   *A Plexsys 45-watt survey transmitter. Photo courtesy of Plexsys.*

output at these frequencies, which was designed for cellular radio, as illustrated in Figure 5.21.

For temporary installations (such as surveys), it is usually much more convenient to use lower-power transmitters (especially if they are being run from batteries) and small antenna feeders that are easier to handle.

The antennas used for surveying are preferably 6 dB. This avoids the need to get the antennas accurately vertical (as would be necessary for higher-gain units) and means that they are physically reasonably small and manageable. Lower-gain antennas reduce the ERP to a point where it might be difficult to measure far-field regions. At lower frequencies, high-gain antennas can be unmanageably large.

Combining all these factors, you will probably find that the survey ERP is about a magnitude lower than a medium power base station. This can be easily corrected mathematically, but places some constraints on the receiver S/N performance because it means that the receiver must be functional at very low received-signal levels.

Figure 5.22 shows a typical configuration of survey equipment. The ERP of the transmitter is (in dB)

$$10 \log T \text{ (watts)} - L + A$$

To convert the field strength measured by this arrangement to an equivalent cellular system level, it is necessary to correct for ERP and

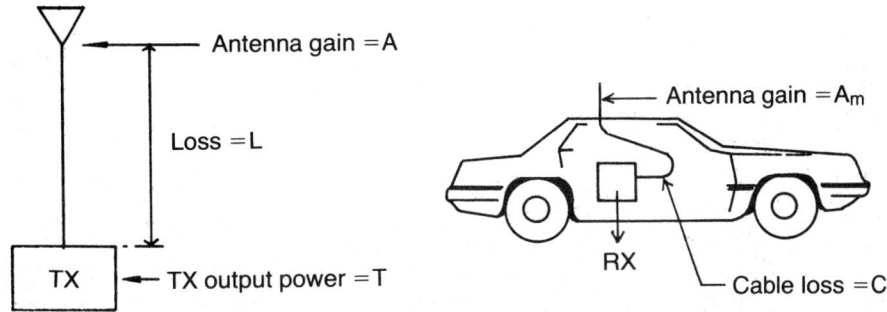

Figure 5.22   *A typical survey equipment configuration.*

receiver gain. Thus, if the cell site has an ERP of E watts, the correction for the transmission system is

$$10 \log E - 10 \log T + L - A$$

Including the passive receive gain, a correction factor of $10 \log E - 10 \log T + L - A - A_m + C$ must be applied.

Generally, the availability of transmitters on frequencies below 500 MHz is much greater than those above that frequency. It is also common for 500 MHz transmitters to have power outputs from 10–25 watts, compared to 800 MHz units, which are usually limited to about 10 watts. In most respects, frequencies from 400–500 MHz accurately enough model propagation at 800–900 MHz if a 2 dBµV/m allowance is made for slightly lower path losses at the lower frequency.

Be cautious if this is done; the 2 dB correction factor applies only to field strengths measured in dBµV/m. In other units (dBm, dBµV, µV), due allowance must be made for the aperture of the antenna.

For survey purposes, the transmitters must be rated for continuous operation; you should note that most two-way radios with PTT (Press To Talk) switches are not so rated. Generally, it is necessary to de-rate the output power of such transmitters to 20 percent if continuous operation is expected. Even then, mounting a computer fan over the heat sink to provide adequate ventilation is a good idea because the difference between continuous and non-continuous rated transmitters relates to the ability of the RF power amplifiers to dissipate heat.

Notice also that continuously rated mobiles are normally duplex and have a duplex coupler to the antenna. This normally has a 3-dB loss (meaning 50 percent of the output power is lost in the duplexer), so the power efficiency is low. This can particularly be a problem when the survey station runs on batteries (as is often the case). The duplexer can be bypassed to increase ERP or save battery power as required.

When batteries are used to power the survey transmitter, it is a good idea to purchase a timer to switch on the transmitter just before work begins in the morning and switch it off in the evening. This technique usually allows three days of operation from a pair of truck batteries and saves many visits to the site. Commercially available domestic power timers with clockwork timers usually serve this purpose adequately and have the advantage of being very cheap.

AUTOMATIC POSITION-LOCATING SYSTEMS

An automatic position-location system records the position as well as the field strength each time a reading is taken. This is particularly valuable for plotting coverage areas by computer. Automatic position-locating systems can usually be described as either too inaccurate (inertial and magnetic), too expensive (omega and high-quality inertial), or too slow (satellite). Table 5.1 illustrates these systems.

Those systems that are inaccurate (inertial and magnetic) usually rely on manual correction every few kilometers. This is time-con-

Table 5.1   *Accuracy and costs of mobile position-location systems.*

SYSTEM	COVERAGE	COST ($US)	TYPICAL ACCURACY (meters)
Loran C	Most parts of the world	2,000	400
Omega	Worldwide	4–20,000	6,000
Aviation	Worldwide	5,000	200*
GPS	Worldwide	2,000	10*
Cellular telephone	Selected areas	2,000	400
Terrestrial mobile	Selected areas	3,000	400
*with correction			

suming and can also introduce significant errors if the correction points get misplaced.

Loran C systems can be used with just-acceptable accuracy and reasonable prices. Omega and high-quality inertial systems, which are sufficiently accurate for survey (accurate to 400 meters), are only practical where the budget is virtually unlimited. Aviation systems are suitable only for airborne applications and are not designed to work terrestrially. The satellite method is limited to periodic fixes every 20 minutes or so. In most parts of the world, however, this method holds the most promise for the future.

The GPS (Global Positioning System) system provides a worldwide 24-hour satellite navigational system that is accurate to a few hundred meters. There are two levels of access defined by the PIN codes of the users so that civilian users can receive to an accuracy of 100 meters but 10-meter resolution is available to military users.

The GPS system works on the principle that it is possible to calculate an exact position in three dimensions if the distance to three reference points is known. The system uses the delay of the transmission from each satellite to determine the distance, and it reads the location of the satellite from the satellite broadcasts.

The range R_i as calculated in Equation 5.1 can be found for each satellite within range. Provided at least three such satellites are within range, the system, to fully determine its location, merely needs to solve the three simultaneous equations.

Equation 5.1

$$R_i^2 = (X - X_{si})^2 + (Y - Y_{si})^2 + (Z - Z_{si})^2$$

where

R_i = Range to satellite i

(determined by propagation delay)

X, Y, Z = Three-dimensional coordinates
of the receiver

X_{si}, Y_{si}, Z_{si} = Three-dimensional coordinates of the satellite
(as broadcast)

It is a little more complex when you realize that the range, as determined by the propagation delay, will have an error proportional to the time error of the reference clock in the mobile. Thus, if the time base of the mobile receiver is ΔT seconds in error, then the error in the range calculation will be $R_E = \Delta T \times C$ (where C = the speed of light).

To avoid the need (and expense) of all receivers carrying atomic clocks, a fourth equation containing the time error can be solved to allow for the inaccuracy of the receiver's clock.

So, the coordinate Equation 5.1 is replaced with the following equation and solved as four simultaneous equations which are now independent of the accuracy of the mobile clock.

Equation 5.2

$$(R_i - R_E)^2 = (X - X_{si})^2 + (Y - Y_{si})^2 + (Z - Z_{si})^2$$

An accurate mobile time base is no longer required.

Since Equation 5.2 has four unknown variables, it requires a fourth equation to solve it. Thus, if reasonably priced timepieces are to be used at the receiving end, it is necessary to simultaneously obtain data from at least four satellites.

The GPS plan is for 24 satellites to be launched with 18 active units in six different orbits plus one spare in each orbit. The satellites are in an inclined orbit which takes them over any point in their path once every 12 hours.

The deployment of these satellites was held up by the delays experienced with the NASA shuttle.

The satellites use spread-spectrum techniques with a signal bandwidth of 2 MHz but an inherent base band of 100 Hz. This enables the system to work down to –163 dBm (a very low power density). The system has a relatively inexpensive receiver (approximately $2,000) and RS 232 output.

Semi-Automatic Position Location

This involves using the survey map with a digitizer board. When a reading is required, the surveyor touches a pen to the map location, which automatically takes a reading and records (accurately) the location coordinates.

Manual Position Location

Unless a very good position-location system is used, it is necessary to use two people in a survey environment. When two people are used, the manual location does not overtax the non-driver; this is both the cheapest and most accurate method.

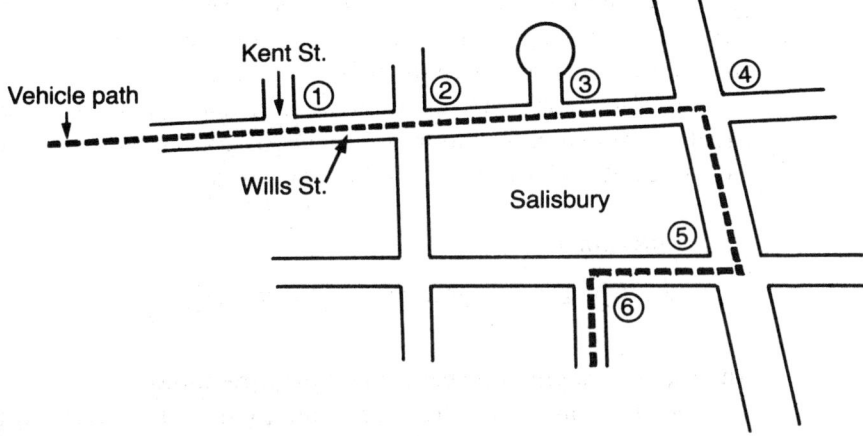

Figure 5.23 *Map identification of samples.*

In normal suburban environments, it is recommended that position fixes be street intersections and that reading numbers be marked on the map in real time. Figure 5.23 illustrates the path of a typical survey run. Samples are taken at the points marked 1, 2, and so on, and the run numbers are marked on the map as the samples are taken. A conventional street map in black and white is preferred, so that the run numbers are easily seen when marked off with red pen. At some later time, the field-strength readings are transferred to the map, using another readily visible color.

PREPARATION OF RESULTS

If the results are manually collected, they should be transferred to a street map. The map should clearly show the following information:

- Date of survey and site surveyed
- ERP of transmitter
- Surveyed frequency
- Antenna gain
- Any correction factors

The run number and field strength should be marked on the map in different colors and it should be clearly indicated which is which. In

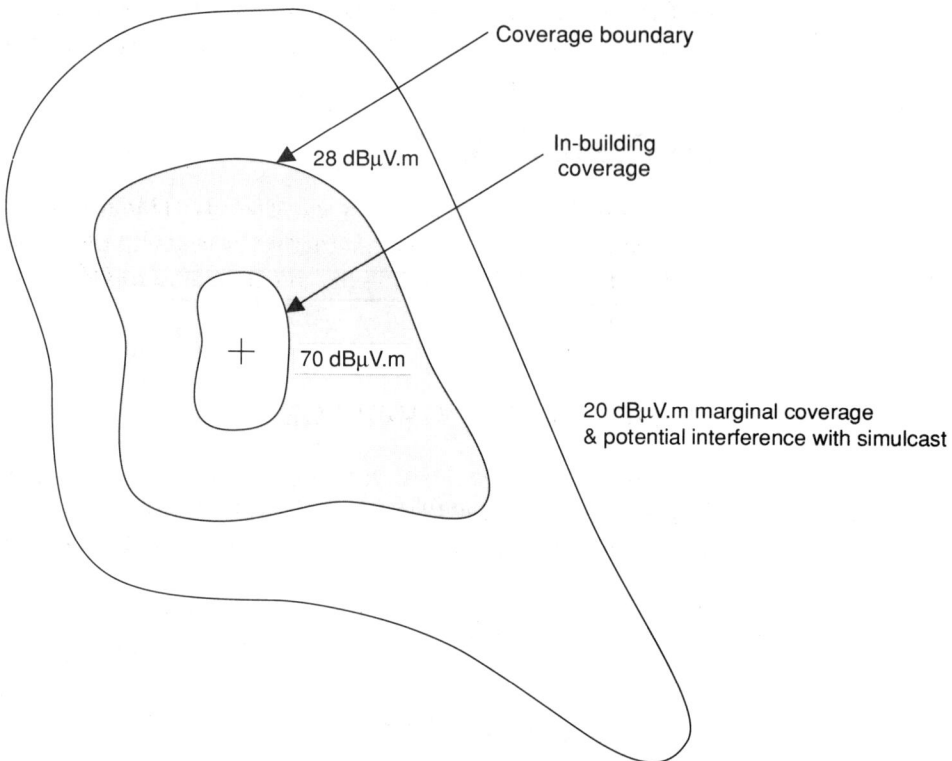

Figure 5.24 *Plotted field strength from survey*

order to make the results easier to visualize, it is necessary to draw the service-area contour (say, 28 dBμV/m) and the CBD contour (70+ dBμV/m).

Because simulcast is usually an important consideration, it is necessary to survey down to the 20 dBμV/m contour (the level at which interference becomes service affecting). See Figure 5.24. This defines the region where equalization of time delays will be required.

In order to see how various sites will work together, it is a good idea to transfer the coverage contour to a sheet of transparent plastic so that you can clearly see the overlapping of different transmitters.

Some map or computer studies will have been done before selecting sites to be surveyed. The survey results should always be compared with the predictions and discrepancies explained. A large discrepancy means either a problem in the forecasting technique or a problem with the survey procedure. The comparison can be done quickly and will frequently highlight problems.

It is advisable to have a second person check all correction factors. Use Table 4.2 in Chapter 4, to assist with this conversion. At higher frequencies (500 MHz+), the cable loss from the mobile antenna to the receiver port is about 3 dB and cannot be neglected.

SPECTRUM CHECK

While the survey antenna is on site, it is advisable to scan the band to be used from the survey site. Other services using the band will then be detected. For example, microwave links and mobiles are occasionally found.

CONFIRMING COVERAGE

For turnkey projects, the operator often specifies the coverage as being at a certain quality for, say, 90 percent of the area for 90 percent of the time. Suppliers often undertake to guarantee the stated coverage.

How can an operator confirm that the supplier's design meets the specification? To do this for a real city is almost impossible, but to see how it might be done, consider the case of an imaginary town, called "*Square Town*," which is shown in Figure 5.25.

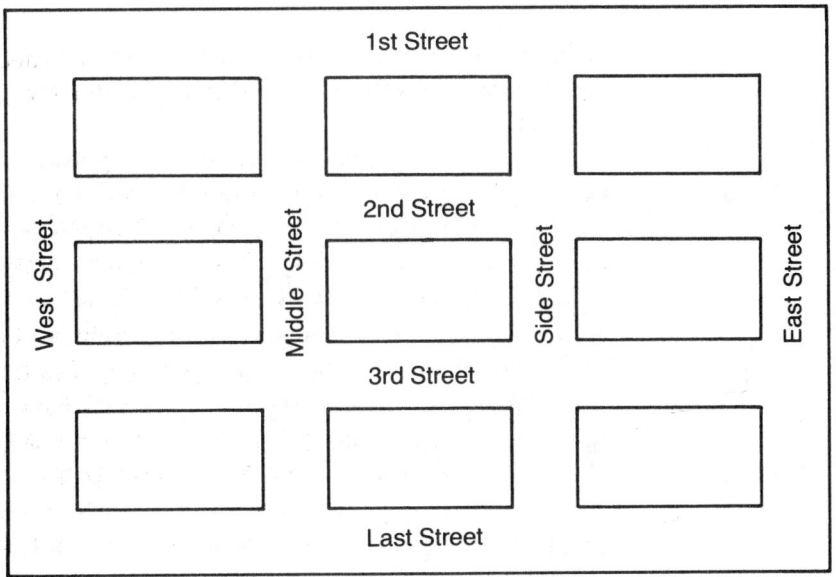

Figure 5.25   *Square Town city map.*

If the operator had specified that the coverage should be such that calls could be reliably received from 90 percent of locations for 90 percent of the time (most unwisely, as this means 10 percent of the area *is totally* unserviceable), then it would be possible to proceed by attempting to receive calls at various locations around the city and determining that the success rate is 90 percent or better. The results would not be conclusive, however, unless every street was sampled at the Nyquist rate (every 0.083 meters). If the town was 2 km × 2 km, then the number of samples would be 200,000, or 8 streets × 2 km/$0.083 × 10^3$ call attempts. Statistical sampling methods could also be used to reduce the sample size (that is explained in the next example).

If the specification called for a minimum field strength—say, F, in 90 percent of locations for 90 percent of the time—then there are two practical ways of measuring this value. First, you could calculate the corresponding mean from the *estimated* standard deviation for a field strength of F at 90 percent of the time/locations and measure the mean field strength (sampling at or about the Nyquist rate). This should be done on a street-by-street basis and, provided 90 percent of all readings are above the calculated average, then the criteria is satisfied. Or, second, you could use a measuring set that can return the field strength above which 90 percent of all readings occur.

The second method is equivalent but more accurate than the first method. The whole city can then be measured as a single entity. If the measurements on this basis yield field strength of F or higher, then the criteria is satisfied.

If a signal-to-noise ratio was specified instead of a field strength, then the measurements above, on a modulated carrier using a S/N measuring set, would be necessary. Such equipment is not commonly available.

The difficulty of performing the above measurements in a real city can be enormous. A real town, for example, might consist of 50 percent developed land and 50 percent rugged hills. An operator might find inadequate coverage in the city and complain that 20 percent of the built-up area (measured as above) is below standard coverage. The supplier could counter claim that all of the undeveloped area is covered at or above the standard level and therefore 90 percent of the city is covered at or above the specified level.

Thus, coverage guarantees for most real cities are not worth much and probably could not be effectively challenged in court. I know of no case where the guarantee has resulted in the guarantor paying any damages or costs associated with unsatisfactory coverage, and yet less-than-expected coverage is a frequent complaint.

Alternatively, the city to be tested can be divided up into a grid, and a sufficient number of coordinates can be selected to satisfy that samples taken in these regions will yield the average field strength to a given degree of confidence. This method has, in fact, been used by a number of system operators, but it has some serious limitations.

First, the sample length should be defined to be, for example, 100 meters, sampling twice every wavelength. The random selection will probably yield points that are not directly measurable (that is, a path diagonally across a city block may be chosen). In order to avoid bias (either positive or negative), it is then necessary to generate a set of unambiguous rules to select the nearest practical path—both its start and end. Next, again to avoid bias, as the interpretation of the selected path would still be quite open (to the survey team), it would be necessary to have some rules that would prevent choosing a biased sample space during measurement. One method might be to increase the sample run to the whole city block. With careful control, this method may be the most practical way of ascertaining coverage standards.

SURVEYING AS A MAINTENANCE TOOL

Very few operators fully appreciate the value of surveying as a maintenance tool. Provided the original design and acceptance-survey results are well documented, the survey equipment can be very effectively used for future maintenance.

No matter how careful the original design/surveys were, in any city of significant size it is not practical to survey every street. Soon after start-up, the operator will probably discover some areas covered that would not have been expected to be, and also some areas not covered for which good coverage was anticipated. Indirectly, therefore, some interpolation/extrapolation of measured result is needed.

When complaints come in (as they will) about coverage, the first reference is the original survey map. When, however, the area from which the complaint comes was not originally surveyed in detail, it may be necessary to look further. A survey vehicle should be sent to the area and a detailed survey of the suspect area taken. Always check nearby areas that were originally surveyed to confirm that the coverage has not deteriorated since the original acceptance survey. If there are discrepancies between the original survey and the check, then a base-station fault can be expected.

Sometimes you will find that adequate field strength is present but that calls still cannot be successfully made. This can probably be traced to simulcast interference. As a matter of routine maintenance, it is a good idea to spot check the field strength from each base station on a yearly basis (or more often if complaints warrant it).

SOME NECESSARY PRECAUTIONS FOR RADIO SURVEY

The following considerations are important precautions for a radio survey: air conditioning, possible errors, and equipment stability.

Air Conditioning

The survey vehicle should be air conditioned to reduce errors due to temperature-sensitive drifts in the receivers and A/D card. Air conditioning also helps the operators stay alert so they are more likely to pick up potential sources of errors.

Errors can occur for a large number of reasons and, where real-time output of results is available, those results should be checked for consistency as the survey progresses.

Possible Errors

Possible errors and their sources are considered in the following list:

- Receiver out of calibration. This should rarely occur with a good receiver but can be quite a problem when cheaper (or older) mobile radios are used as the measuring device. Frequent checks (weekly on suspect receivers and monthly on quality measuring receivers) of calibration against a good signal generator are necessary.
- Faulty antenna. The antenna should be free of visible defects and have a low VSWR. It should occasionally be checked against a few other similar antennas to confirm gain, using different feeders to the receiver.
- Faulty or damaged feeders. Use the same checks as for antennas.

- Mobile receiver de-tuned or simply not tuned to correct frequency. Double-check that the receiver is tuned to the correct frequency. The mobile can also be desensitized by other transmitting apparatus in the vehicle, particularly in regions of low field strength. Beware of transposing the TX and RX channels when setting the frequency.

- Wrong calibration table used. This often becomes a problem when the units of the signal generator output are not the same as the units used for field strength (for example, the signal generator is calibrated in dBμ and the field strength equipment is calibrated in dBμV/m). Converting units in the field often results in error. Always have a correction table handy (for example, use Table 4.2, found in Chapter 4).

- Test-receiver output is voltage/temperature dependent. Always check the test receiver for temperature and voltage sensitivity. This is more likely to be a problem with old receivers, but all sets should be checked before being placed into service.
  The voltage regulation on a car battery may not be good, particularly if the receiver gets its power from a source distant from the battery where significant voltage drops may occur. Be cautious of volt drops from indicators and brake lights.

- Insufficient settling time. Even good quality measuring sets should be powered up half an hour before beginning measuring. Most of the drift occurs in the first 10 minutes after the set is switched on.

- Inaccurate records of transmitter base. Keep good records of the survey conditions, particularly for a temporary test transmitter. In particular, record the following:

  1. Power output at start (measured)
  2. Feeder loss (cable should be calibrated)
  3. Antenna gain, antenna height
  4. Frequency of test transmitter
  5. Power at end of test (remeasure to ensure no drift has occurred)
  6. Date, and TX site name

Failure to record any of these details could render the data useless in the future.

Equipment Stability

It is most important that all items in a moving vehicle be securely fixed. It is usually necessary to mount the receiver and other survey hardware in a rack. This rack should be padded and constructed in such a way that it does not interfere with the vision of the driver (this usually limits the rack height).

It is best if the equipment can be mounted beside the operator, where, in the event of an accident, it is unlikely to come in contact with the operator. The most dangerous mounting position is in front of the operator, where a collision will throw the operator into it. This is particularly true of a VDU.

An internal master switch for battery power also should be provided, as should a fire extinguisher (CO_2 type). For security, it is best if the rack can be completely covered, in a way to hide the hardware it contains. When it is necessary to leave the vehicle parked in the street for some time, the sight of a few expensive measuring receivers and some comwuter hardware is likely to attract unwanted attention.

Where possible, use laptop computers. Because they are much smaller, laptops can be placed safely on the operator's lap, requiring only an up-and-down movement of the head. An equipment rack mounted at the side of the operator places some strain on the operator's neck muscles when he or she is viewing the keyboard or screen. The operator must a have clear field of vision forward to allow for proper street identification and to minimize the effect of car-sickness. The motion sickness that results from operating survey equipment is similar to that caused by reading in a moving vehicle. It is such a problem for some people that they are unable to do this task effectively.

CHAPTER

6

 # SITE SELECTION

Site selection for paging transmitters is basically a search for good "broadcast" sites. In most towns and cities there will be a well-established mobile-radio network, and the sites chosen for good mobile propagation will generally prove to be suitable for paging. Like mobile radio, paging transmitters consume only modest power (typically a few hundred watts) and one rack is usually adequate to mount the hardware. Because of this, it is not particularly difficult to find suitable accommodation in rented premises.

Space requirements for paging are minimal. Usually the equipment will needs only one 19-inch rack space, and the battery backup of a few hundred ampere hours can be incorporated in the same rack. If a separate link to the controller is used (such as a single channel transceiver) an additional half a rack space may be necessary.

The antenna is relatively small, typically being a collinear device less than 6 m long, which needs to be mounted on a roof-top, it will weigh only a few kilograms. If a single-channel link is provided, its antenna may be as large as the paging transmitter antenna.

Air conditioning, although desirable, is not essential for most base sites but may be required for some models of transmitters.

Although broadcast-type sites are a natural choice for paging operators, an allowance must be made for the business user who will expect good service in the heart of the central business district (CBD).

As a general rule for a city with a population greater than 50,000, it will be necessary to place a transmitter in the central area, not more than 1 km from the geographic center, to ensure good penetration of the high-rise buildings. In bigger cities' high-density areas (where most buildings are higher than two floors, including the ground floor), you should arrange to have local transmitters placed so that no substantial part of the high-rise area is more than 2 km from a transmitter.

Paging transmitters need links back to the controller. You must ensure that the site chosen either has conventional PSTN lines or is suitably placed to enable a microwave link to be installed. Links can be a relatively costly part of the paging installation, and this cost can increase dramatically if their provision is not considered at the time the base-station site is selected.

Paging transmitters themselves are not particularly subject to interference problems, but they are prone to causing interference to other services operating from the same site. The relatively high ERP and bursty nature of the transmissions can cause annoying interference to others. Community sites that may be suitable for paging are not always particularly well managed, and often they come with built-in intermodulation hazards like rusty fences, antennas seemingly carelessly placed, and mixtures of new and antiquated hardware. Sites that are bristling with antennas, and particularly ones that look a little run down, can cause serious problems for the paging operator. As a rule, if the new transmitter causes interference to another user who was on site first, it is "accepted" that the newcomer has to pay to rectify the situation. Often the problem may be traced to substandard equipment used by the complainant, but it will be the pager operator who will have to spend the time to identify the problem. As this is such a time-consuming (and hence costly) process, and in the interest of good long-term relationships with the other users, it generally is a wise policy to spend a little more and pay for the rectification.

DEFINING COVERAGE

Coverage is usually defined to be that area for which coverage is reliable to the extent that 90 percent of the area is covered for 90 percent of the time in the fringe areas. It is important to distinguish between coverage defined as 90 percent of the area 90 percent and time (which means 10 percent of all calls fail), and the earlier definition, which meant that this high call-failure rate appeared only at the boundary and in restricted fringe coverage areas. Generally, within the areas covered, the call-success rate will be much higher than 90 percent and within the coverage area will be better than 97 percent.

Sometimes the system designer will be confronted with the situation where management tries to "define" the coverage area before any map studies or surveys are undertaken. Such definitions usually do not account for the nature of radio propagation and assume that sharp boundaries can be "tailored" and that hills pose no problems. While to some extent the designer can alter the coverage pattern, it is the topography that is the strongest determinant of coverage, and it is essential that at an early stage, management and engineering discuss what is economically possible and what is desired.

CENTRAL BUSINESS DISTRICTS (CBD)

It is virtually impossible to perform any meaningful survey or map forecast for a modern, high-rise central business district. Although assumptions may be made about high-rise buildings, such as attenuation by the structure will be between 20 and 30 dB, it must be remembered that these assumptions are quite vague. If the street level RF is 30 dB above the pager design level, then it is probably safe to assume that reasonable coverage can be obtained within the building.

A "rule of thumb" for CBD coverage is that all buildings in the service area should be no more that 1.5 km from a base station (assuming 100-watt ERP base stations).

When designing citywide coverage it is generally best to start by defining the location of the CBD transmitters and then measuring (or otherwise determining) the coverage from these sites. This is because acquisition of suitable CBD sites can be tedious and time consuming, and once a reasonable site has been located it is best to use it rather than to waste too much time with optimization, which inevitably leads to the conclusion that the sites that can't be acquired are better. Once the CBD sites have been finalized, it is time to select suburban sites to provide contiguous coverage. In the suburbs there will be more choice and flexibility on site location.

Standby or Site Redundancy

With paging systems there are two practical types of redundancy. Conventional redundancy involves the use of parallel equipment installation so that the failure of one piece of hardware is backed up by the standby. Because paging base-station equipment is not overly expensive and the footprint is moderate, an alternative worth considering is to have stand-alone base sites but provide redundancy by placing additional sites which overlap sufficiently that the failure of one or more transmitters will not affect service. In large systems this can provide both a better alternative economically and better coverage to some of the more difficult areas from more than one base, as seen in Figure 6.1.

Suburban and Rural Areas

In the suburbs and rural areas it is relatively easy to quantify coverage on the basis of field strength. Depending to some extent on the actual pager receivers used, a design field strength of 28–32 dBµV/m will usually be adequate to define the boundary of the coverage.

MAP STUDIES

Preliminary design can be done by map study or computer-aided design. Computer design is more common today, and over-reliance on computer predictions is a major source of coverage problems. Neither map studies nor computer aided techniques are as accurate as a well-conducted survey. You should beware also that the most expensive computer packages are not always the most accurate.

Reasonably accurate predictions of coverage can be made using the techniques found in the following reports.

- CCIR, "Recommendations and Reports of the CCIR," 1982, Volume V, Propagation in Non-Ionised Media, Report 567-2.
- Yoshihisa Okumura et al., *Review of the Electrical Communication Laboratory*, Volume 16, Nos. 9–10 September–October, 1968. NTT, Japan.

These methods are empirical but are well tried and proven. They are the results of a series of trials based on transmitters sited at different elevations and terrains, and the resultant field strength.

While both of the above methods are equally effective, the Okumura method may be more useful for paging as it specifically covers 150, 450, 900, and 1500 MHz, while the CCIR report covers only 450 and 900 MHz.

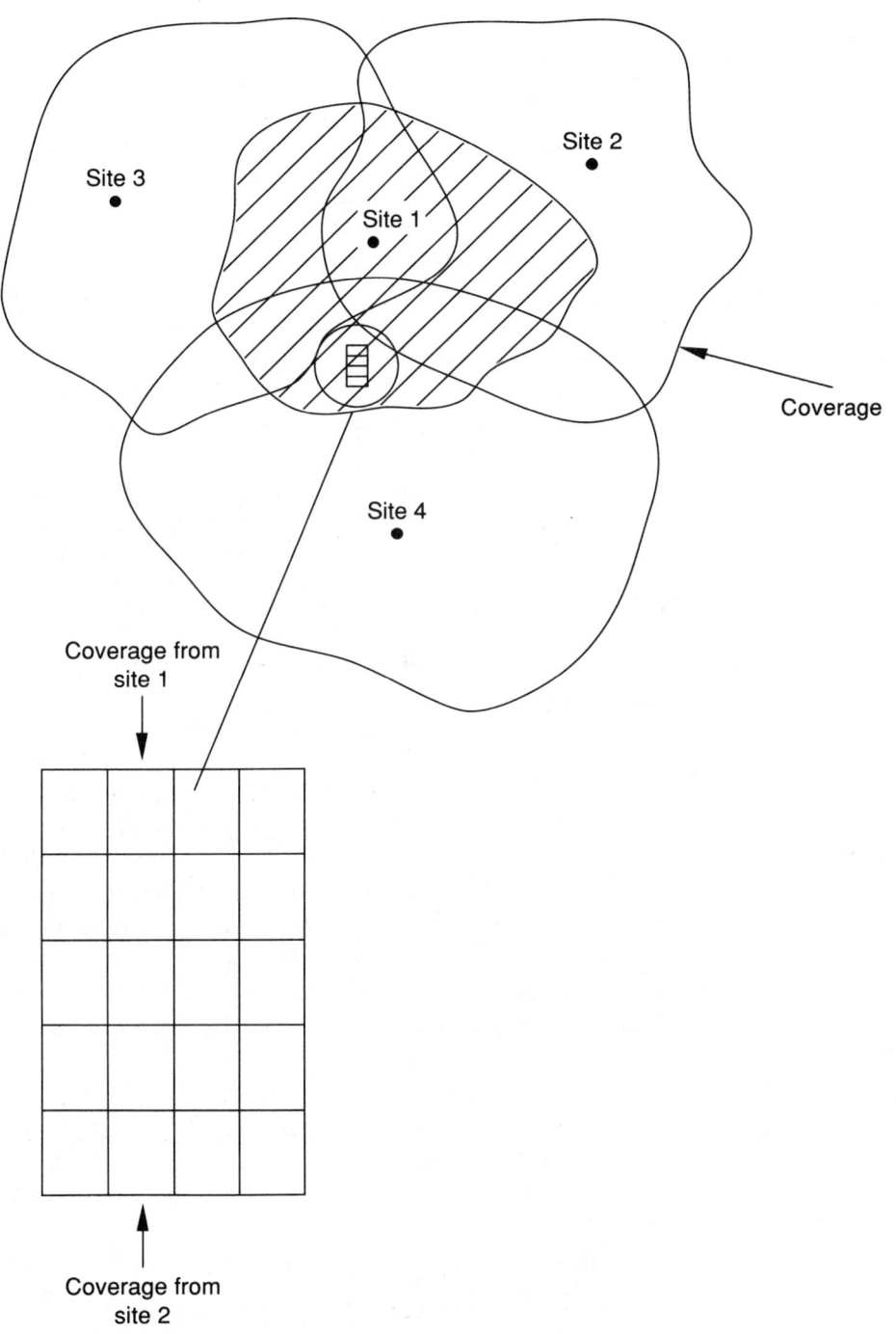

Figure 6.1   *Redundancy with overlapping coverage.*

COMPUTERIZED TECHNIQUES

Computerized methods using digitized maps can produce a higher degree of accuracy but are very costly. The accuracy of good computer techniques is still only ±6 dB (compared to ±10 dB for manual methods).

Computer prediction methods normally assume a two-dimensional path between the transmitter and receiver. In real-life propagation, however, contributions to the far-field pattern are made by reflected, scattered, and refracted paths that are in other planes. Wave scattering produces a spatial spectrum that is a complex holographic image of the surface causing the scattering. It is generally not practical to take these effects into account. Therefore, there is a limit to the degree to which calculations can reflect reality.

The detail needed to accurately determine the path profiles is also very large. If a city of 2000 km^2 is characterized by 100m × 100m areas, then there are 2000 × 10 × 10, or 200,000 such areas. A minimum representation would contain two pieces of information about each area, namely height above mean sea level and type of terrain (that is, urban, rural, open, or water), and thus would contain a total of 400,000 pieces of information. It is also desirable to store information about surface clutter (whether man-made, like buildings, or natural, like trees) and its height and distribution. This information can be the most difficult to obtain and to update. Moreover, for computer systems to be useful, they must not be limited to one city.

Thus, a large digital database and data storage in the high-megabyte range is necessary. The acquisition of this data is costly and is, ultimately, the limiting factor. Unless a database is available from other sources (for example, mapping authorities), the cost of producing one may be too high to be undertaken by a cellular operator. However, most computer-prediction databases lack surface-clutter information and are of limited use.

Despite the apparent limitations, computerized techniques have found favor among many designers, and most major operators and suppliers have an in-house system (which has generally been customized). These systems can generally do more than just forecast propagation. Most can graphically depict composite coverage, which has been derived from the files of a number of individual site predictions. As an extension of this, composite interference maps can be produced. These can be particularly useful for gaining an overview of a system's performance. A number of these forecasting systems are available for sale. Typically, the cost will be about $100,000 to

$200,000. Additionally, hardware consisting of a minicomputer, color monitor, one gigabyte-plus storage, and a large, color graphics printer is needed.

There are a number of companies offering computer prediction services at very reasonable rates (in the vicinity of $200 per plot). These simply require the operator to specify the base-station height, power, and coordinates. Of course it is assumed that the operator is in a region for which digitized maps are readily available. If they are not, the cost of producing them would be prohibitive. For small operators an outside computer prediction would be a good alternative to an in-house system.

MANUAL PROPAGATION PREDICTION

A number of empirical studies have produced algorithms that can be used to determine the far-field strength as a function of ERP (Effective Radiated Power) and range. Generally these results were produced after an extensive series of propagation tests in one or, at most, a few countries.

Studies done in different countries (with different terrains) revealed substantial variations in path attenuation over similar terrain. One of the main sources of discrepancy is in the description of topography. A hill in Venice may be compared to a molehill in San Francisco!

Thus, if a standard series of published curves (for example, as in Figure 6.2) are used to determine range, it is often wise to first "calibrate" the curves by measurement. In this process, a field-strength survey is undertaken to enable a detailed comparison between reality and the model. If a correction of more than 3 dB is necessary, additional measurements should be taken in various terrains to improve the accuracy.

ESTIMATING BASE-STATION RANGE

Let's assume you want to estimate the range for a typical base site; let's also assume a 20-watt TX transmitter power at 900 MHz. Proceed as follows:

1. Correct for actual ERP (Effective Radiated Power).

    The graph is drawn for 1000-watt ERP and so a suitable correction factor needs to be applied for the ERP actually being trans-

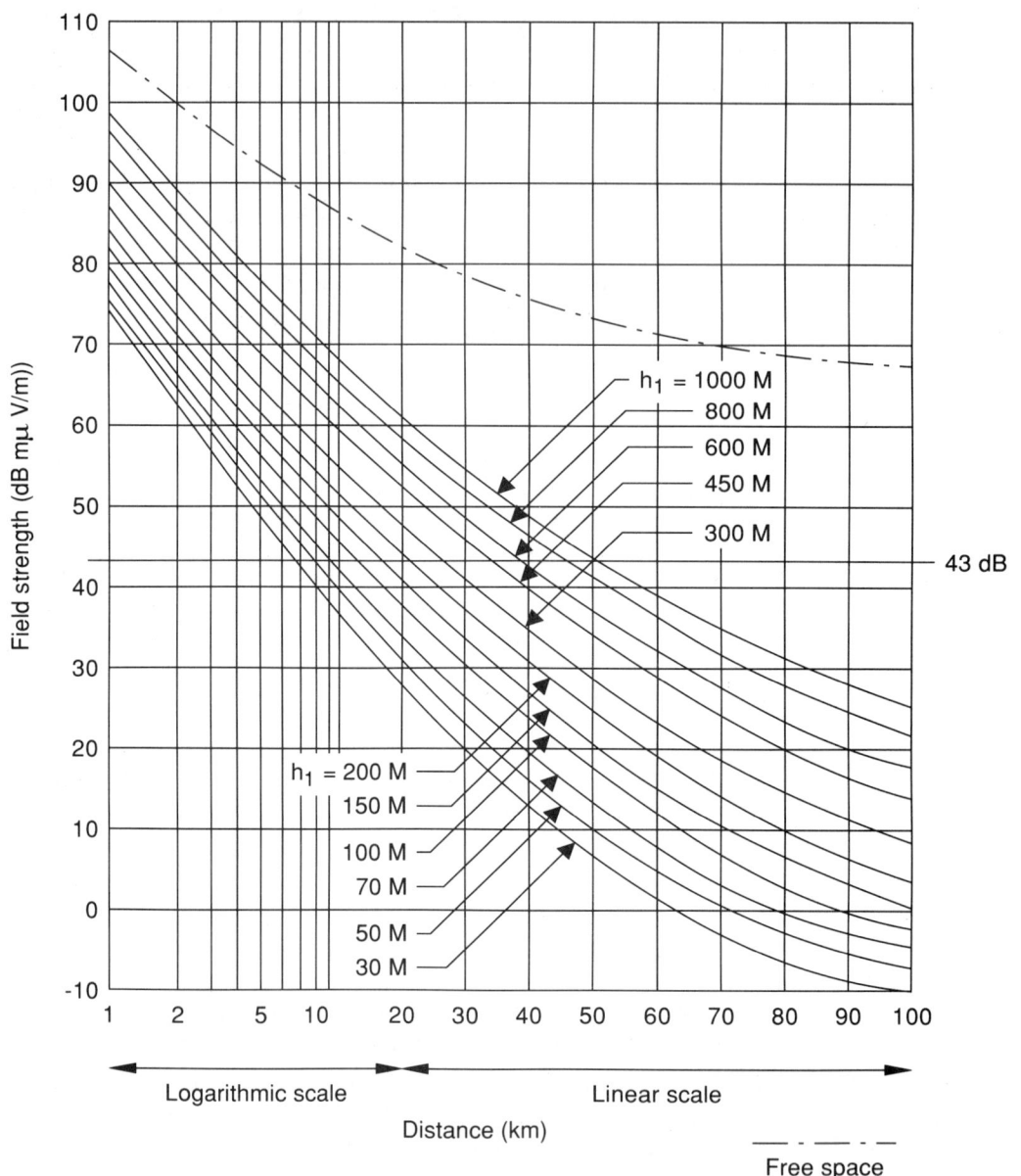

Frequency ≈ 900 MHz; Urban area 50% of the time;
50% of the locations; h_2 = 1.5 m

Figure 6.2   *Field strength (dBμV/m for 1 km ERP) in an urban area.*

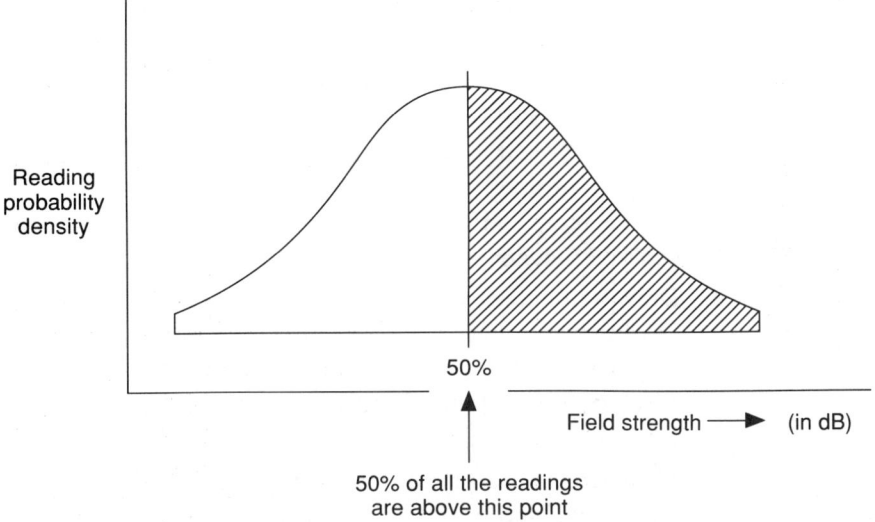

Figure 6.3   *Field strength is log-normally distributed. The shaded area represents the probability that field strength will exceed X. The value thus obtained is really the median value of the log of the field strength and will differ from the actual mean field strength. In practice, this discrepancy is generally ignored.*

mitted. The ERP of a 20-watt TX is about 50 watts (depending on feeder loss, antenna gain, etc).

$$\text{Therefore Factor} = + 10 \log \frac{1000}{50}$$

$$= + 13 \text{ dB}$$

Thus, the 30-dBµV/m boundary (for AMPS) will be located on this curve at 30 + 13 = 43 dBµV/m.

2. Draw the graphs for the field strength attained at 50 percent of locations and times; that is, the field strength that is exceeded in 50 percent of all readings. Field strength is a log-normally distributed variable and is illustrated in Figure 6.3.

    If the standard deviation "σ" of the field strength is not known, it can be characterized as urban (σ = 8 dB to 12 dB), suburban (σ = 6 dB), rural (σ = 3 dB), or water paths (σ = 1.5 dB). Some propagation graphs show the field strength for 90 percent of locations and time. This can similarly be corrected to the mean using the relationship mean = (90 percent locations and time reading) + 1.28 × σ.

Thus, to find the distance at which the mean is 30 dB using such a graph, it will be necessary to look for the value of the field strength on the graph corresponding to 30 − 1.28 × σ. If σ = 6 dB, then 30 dBμV/m average = 22.3 dBμV/m on the 90 percent/90 percent graph.

3. Draw a line through 43 dBμV/m on the graph to obtain the 30 dBμV/m contour. You can now obtain the range in an urban environment by reading off the range against this line. A correction must then be made for the actual environment (unless it is urban).

   The graph is drawn for urban environments defined as 15 percent of the land occupied by buildings. In very dense CBDs, this percentage can be much higher, and in rural areas it can be zero. However, the formula is only valid to 2-percent occupancy. A correction factor, $S = 30 − 25 \log \alpha$, can be used, where α equals percent of building to total land area. This percentage can be applied to terrains different from suburban terrains.

   If a correction factor is applied, the new level in Figure 6.3 for 30 dBμV/m (50 percent, 50 percent) is (30 − S). Notice that at α = 15 percent (that is, in an urban environment), S = 0 so that S may take positive or negative values.

   Finer distinctions between terrain types are offered by the Okumura paper but because this whole technique is, at best, approximate, additional corrections are normally not needed except for very unusual terrain. Such terrain includes very flat land (reclaimed swamps, subtract 27 dB or S = 27), water or part water paths (see Okumura), and large hills (treat as absolute boundaries).

4. Find the curve corresponding to the base-station height (h_1) with respect to the surrounding terrain.

5. Read off the range corresponding to the type of terrain; for example, Urban terrain, base-station height h_1 = 30 m, range = 8 km.

6. Plot this coverage on the map, making adjustments for local terrain; for example, hills form natural boundaries.

7. Plot the 30-dBμV/m (50 percent, 50 percent) contour (which is 43 dBμV/m on the graph) or other, as applicable.

Choose sites that look likely to provide good continuous coverage with respect to the central site and then survey them. Note that this process is not used to select sites. It only determines which sites should be surveyed. From the survey results, it will become evident which sites are useful and which are not. The process is then repeated.

Variable Terrain

Where the terrain is variable within one cell coverage area, the plot of the coverage should be done in sectors. Consider the prediction of the coverage for the area in Figure 6.4. There are three discrete types of terrain seen by the cell site. The propagation over each type will be quite different. If the CBD is substantial and the cell site is more than 4 km away, it should be regarded as an absolute barrier.

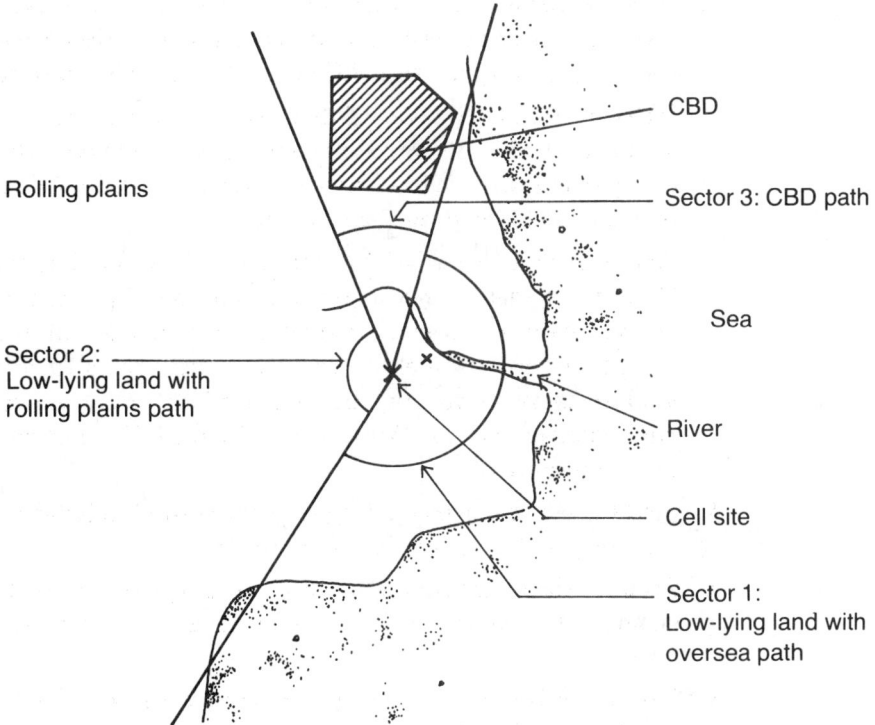

Figure 6.4   *Dealing with variable terrain.*

Sea Paths

Sea paths offer the best radio paths. As boats move slowly (compared to wavelengths/second), they suffer very little from multipath noise. These sea paths can provide any given grade of service (measured as S/N) at a lower field strength than land paths.

A field strength of 15 dBμV/m less than the land path can be regarded as adequate over sea, although a usable service is available at 15 dBμV/m down and sometimes even lower (depending on antenna height). When at sea, the most prominent (highest) land site will be the one with the highest field strength rather than the one closest to the boat. Ranges of 50 km (over unobstructed paths) are typical and up to 100 km are not uncommon under favorable circumstances.

ESTIMATING THE EFFECT OF BASE-STATION HEIGHT

Often it is not possible to get survey results at the desired height. This may be because the original survey was done at a height other than the one chosen for the final installation; or in rural cases it may be that no survey mast is available that equals the height as the proposed tower, or you may simply want to estimate the effect of reducing the height of an existing antenna.

A simple relationship that applies fairly well in practice is that the field strength correction factor depends on height as

$$FS = 20 \times \log (\text{new height/survey height})$$

It can readily be seen that doubling or halving the antenna height will require a correction of 6 dB to *all* the measurements.

Remember that the height in question is the height above ground level (and not simply the tower height) and that if extreme height changes are made, local obstructions may come into play and invalidate the rule. Where there is doubt, resurvey.

SURVEY PLOTS

Survey results are only as good as the equipment and techniques used. Therefore, large discrepancies should ring alarm bells that perhaps something is wrong with your equipment. Assuming that satis-

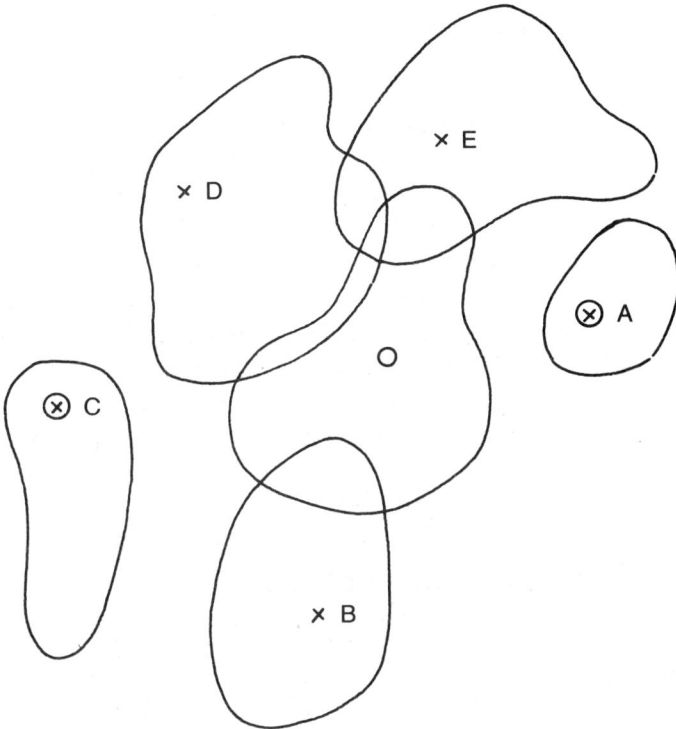

x : Suggested site with potential

ⓧ : Site not suitable

Figure 6.5 *Survey results plotted.*

factory explanations for any discrepancies can be found, the selection process is continued by selecting more sites to provide the remaining coverage, conducting more map (computer) studies and re-surveying until adequate coverage has been found.

Figure 6.5 shows that sites B, D, and E look promising but sites A and C are inadequate. However, before totally dismissing sites A and B, the large discrepancy between the predicted and actual coverage should be examined and the reason for the discrepancy identified.

Because continuous coverage is important, the design proceeds much like putting together a jigsaw puzzle—starting from a fixed point and working outwards.

Site selection involves compromise, particularly with site availability. Fortunately, the designer can tailor coverage by using height, power, and antenna patterns to achieve efficient spectrum reuse.

FILING SURVEYS

The design can be seen to proceed iteratively. Therefore, it is unwise to discard any survey result as totally unsuitable. Unused survey results can be useful in filling gaps in the current or future design. Conversely, as the design evolves, sites that once appeared ideal may become redundant. Thus, all survey results have the potential to become part of the jigsaw puzzle.

When filing surveys, remember that all surveys should be conducted using the same units (dBμV/m is recommended. See Chapter 4, "Units and Concepts of Field Strength," for more information). This will help you avoid confusing results. In addition, all survey maps should be filed with the following data clearly marked:

- Date of survey (very important) and name of surveyor
- Location of site, owner, name, and phone number of contact person
- Survey antenna height (above ground level and above sea level)
- Cable loss (preferably measured because connector losses are difficult to calculate)
- Transmitter power
- Antenna gain (usually 6 dB)
- Correction factor used to convert actual ERP to nominal ERP (that is, all readings are corrected to the ERP of an actual cellular base station)

These details may prove essential in the future if discrepancies are detected.

MAPS AND MAP TABLES

The maps used to plot survey results should be about 1:100,000, certainly not larger than 1:250,000 or smaller than 1:50,000. Topographic projections should be used for coverage prediction. However, street maps that show street names in detail are usually best for recording individual survey results, before they are plotted at a scale suitable for system studies. Actual coverage plotted on street maps is useful for recording detail coverage of individual bases. The final result should be an easy-to-read map designed for use by the subscriber.

A map table of at least 2m × 2m (or larger) is needed to adequately handle coverage maps, given the scale needed to ensure sufficient detail for survey results. Once a large number of maps are stored, it is essential to have a suitable cataloging system. Maps also need adequate storage and a map file (a file that can store full-size maps without folding) is best.

Transparencies can be used to completely overlay the maps to depict coverage. Felt pens used to mark the transparencies should be of the "whiteboard" erasable type. A very handy accessory is a set of map weights. Without weights, maps can be awkward to handle.

CHAPTER

7

❖ ANTENNAS

Generally with paging systems, the aim is to get the highest gain possible from the antenna consistent with the ERP permitted by the regulatory authority. This means mounting it as high as possible, using high-gain antennas and low-loss feeders and connectors.

At the most common paging frequencies, 6-dBd antennas are readily and cheaply available and are the ones most used. At frequencies in the mid-VHF region, 6 dB is about the maximum gain available that is consistent with a reasonable physical size, and equally importantly, adequate stiffness. The gain of a collinear antenna increases directly in proportion to its size. In the mobile field, gain is usually expressed as the log of the gain relative to a half-wave dipole and denoted dBd. For example, at 150 MHz a half-wave dipole is $1/2 \times 300/150$ meters or 1 meter. To get a 3 dBd antenna (3 dB represents double the gain) all that is necessary is to place two dipoles in series, properly phased and separated (see Figure 7.1). Such an antenna would be about 3 m long. To increase the gain to 6 dBd, it is necessary to have four sets of dipoles and the total length would become 6 m. To increase the gain to 9 dB would require a massive 12 m of antenna. Apart from the unwieldy size, such an antenna would have problems with stiffness and would be inclined to sway in the wind. When the narrow vertical patterns of a high-gain antenna are considered, it can be seen that the gain of the antenna would vary with the tilt induced

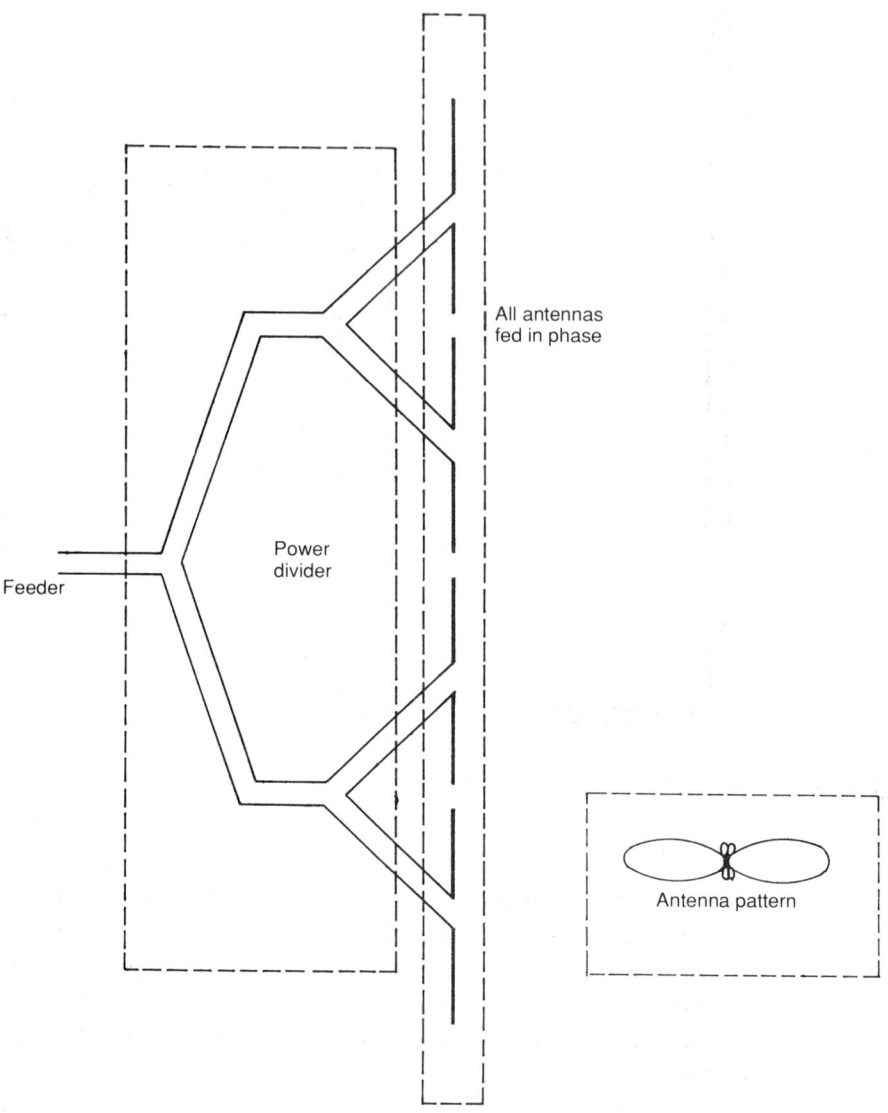

Figure 7.1   *The construction of a simple collinear antenna.*

by the whipping of the antenna in a breeze. Naturally antenna size is directly proportional to the wavelength and so higher gains are practical at higher frequencies. Above 450 MHz, 7.5 dB and higher gains will be available, and up to 12 dBd may be available from some manufacturers at 900 MHz.

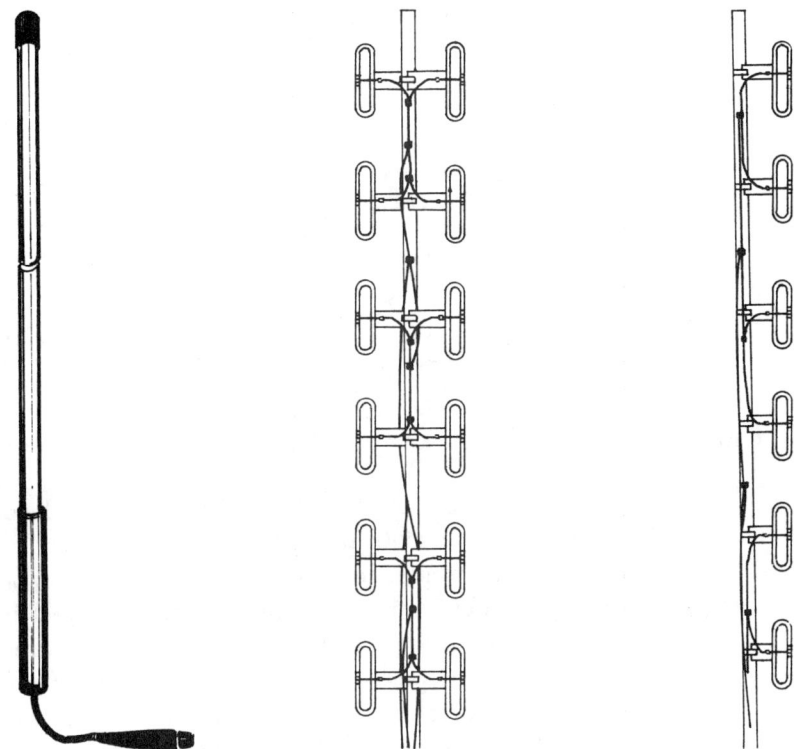

Figure 7.2   *Omni antennas can take all these forms.*

By far the cheapest antenna for paging is the fiberglass encapsulated collinear antenna. These antennas are light and easy to install. This antenna, however, is very susceptible to lightning, and mounting it in such a way that it is within the protective cone of a lightning arrestor can interfere with its radiation pattern. This type of antenna should be mounted well clear of the tower or structure to which it is attached.

A more rugged antenna and one with a degree of built in lightning resistance is the offset dipole. This antenna has a gain similar to that of a collinear of the same size, but is much heavier and costs at least twice as much. This antenna is illustrated in Figure 7.2 along with the more common collinear dipole.

EFFECTIVE RADIATED POWER (ERP)

Effective radiated power is the power that is transmitted in the direction of maximum antenna gain. The ERP of a paging transmitter is

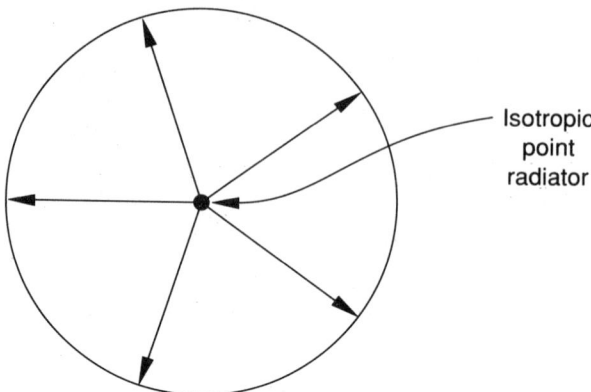

Figure 7.3   *For an isotropic antenna, the radiated power is equal in all directions.*

usually quoted referenced to a dipole, although sometimes the reference will be an isotropic radiator.

An isotropic radiator is one that radiates *equally* in all directions (in a manner similar to a light bulb), as shown in Figure 7.3. For a given input power, the total power radiated from an antenna is constant, but it is possible to shape the radiation pattern so that more power is radiated in some preferred direction. The ratio of the power radiated in the direction of maximum gain of an antenna compared to the power radiated by the reference antenna in its direction of maximum gain is known as the antenna gain.

The dipole antenna has a gain of a little more than 2 dB when compared to an isotropic antenna. The typical radiation of this antenna is shown in Figure 7.4. It can be seen that the radiation pattern is no longer uniform and the compression of the field which results in gain in one direction is counter-balanced by reduced gain (and even nulls) in some directions.

Still higher gains are achievable by further compressing the field patterns in the desired directions.

ANTENNA ARRAYS

At lower frequencies (below 500 MHz), as has already been stated, it is not practical to make antennas with a gain much higher than 6 dB. There are, however, occasions when even at these lower frequencies higher gains would desirable. This can be achieved by the use of simple antenna arrays.

The principle of an array is that two or more antennas are arranged so that the gains of the antennas add constructively in cer-

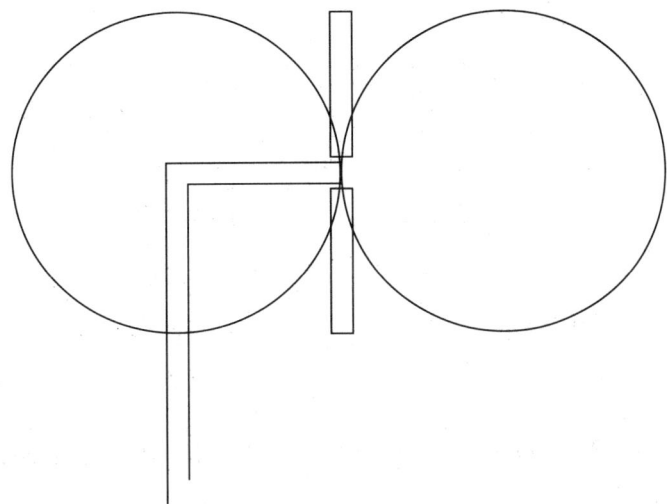

Figure 7.4   *The field pattern of a dipole antenna.*

tain desired directions to produce gains higher than that of either antenna. Conversely the same configuration can be used to produce nulls in certain desired directions.

Generally, the improvements achieved by this method will be limited to gains of about 9 dB maximum, due to higher gains placing very exacting mechanical constraints on the array assembly.

A simple but often effective array consists of two collinear 6-dB antennas connected by a phasing harness and matching transformer. Figure 7.5 shows some of the patterns that can be selected. In each case the maximum gain is 3 dB in the direction indicated.

It is worth noting that a 3-dB gain will not in most cases make a significant difference to coverage, but where areas are already marginally covered the 3 dB so obtained can result in a significant increase in successful pages.

POWER DIVIDER

In order to successfully connect two antennas in parallel, it is important that they be properly matched. Failure to do so will result in losses that are comparable with the gains sought. The simplest impedance matching device is a quarter-wave transformer, connected as shown in Figure 7.6. This transformer has an impedance transfer characteristic such that $Z_{out} \times Z_{in} = Z_{02}$. Cable of 35-ohm impedance is produced for this application.

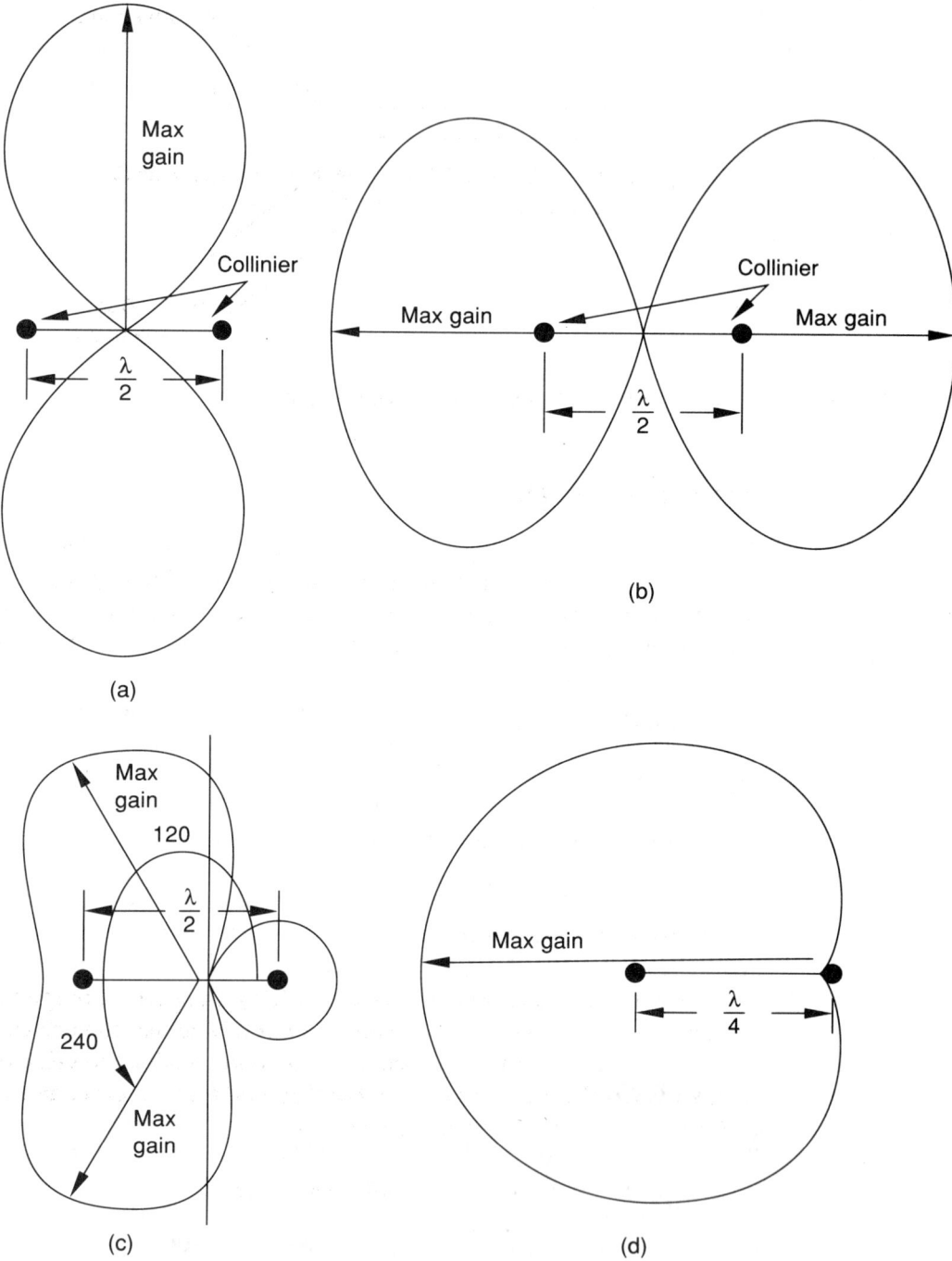

Figure 7.5   *Field patterns with various dipole configurations. (a) Dipoles separated by half a wavelength and fed in phase. (b) Dipoles separated by half a wave-length and fed 180 degrees out of phase. (c) Dipoles separated by half a wavelength and fed 90 degrees out of phase. (d) Dipoles separated by a quarter wavelength and fed 90 degrees out of phase.*

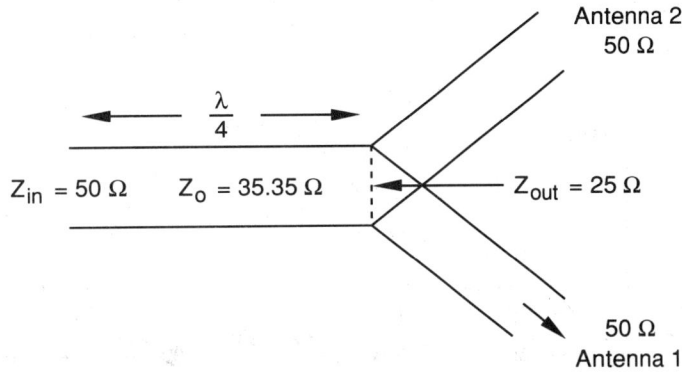

Figure 7.6 *A simple but effective quarter-wave transformer.*

PHASE SHIFTING

It is most important that antenna arrays are phased so that they produce the maximum gain in the desired direction(s). The signal phase is shifted by connecting an additional length of coax cable to one of the antennas to produce the required delay. The delay in wavelengths can be calculated as

$$\text{Delay} = V \times 300/f$$

where

f = frequency in MHz

V = velocity factor

Delay in wavelengths

The velocity factor is the factor by which the speed of propagation in the cable is less than the speed of light. Note that the velocity of light is constant only in a vacuum. In all other media, the velocity is reduced by the dielectric constant of that media. For a few common cables this reduction, or velocity factor, is:

Cable	Velocity Factor
RG213/U	0.66
RG8/U	0.66
RG8/U (foam)	0.80
1/2 inch (foam)	0.88

The cable used for delay should be of the same type as the antenna tail, but more importantly of the same impedance (50 ohms). If possible, extend the tail rather than add a joint, which would introduce losses. It is important to ensure that the delay is added to the correct antenna or the pattern will be reversed.

FEEDERS

The cable feeding the antenna can be a significant source of power loss unless the size is carefully chosen. Bigger cables have lower losses, but they are more costly and difficult to work with. As a guide it is best to keep these losses below 1.5 dB. Cable losses increase with length and frequency as can be seen in Table 7.1.

By far the most popular choice for paging feeder applications is foam-filled heliax cable. This cable is flexible and can be bent into smaller radii turns than solid cables. Although it will have slightly higher losses, it is cheaper than air-filled cable, and does not require the expensive pressurization that is needed in those cables to keep out moisture.

CONNECTORS

Connectors can be a major source of losses, and poor quality connectors can be false economy. Cheap connectors can often test OK for loss when new, but poorly mating contacts or internal corrosion can cause substantial losses. These losses are often the cause of intermodulation problems, which can cause spurious emissions.

Table 7.1   *The length in meters of different cable types for various frequencies to give a maximum 1.5 dB loss.*

Frequency (MHz)	150	450	900
CABLE			
RG213	15	10	5
1/2 inch	45	27	18
7/8 inch	90	45	32
1-1/4 inch	120	65	45

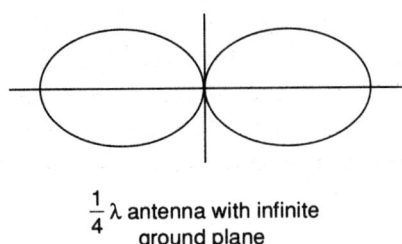

$\frac{1}{4}\lambda$ antenna with infinite
ground plane

Figure 7.7 *A quarter-wave antenna.*

ANTENNA IMPEDANCES

A quarter-wave antenna, as depicted in Figure 7.7, has an effective radiation resistance of 37 ohms when provided with a good ground plane. In this case the radiation pattern will be the same as a dipole. However as the ground plane is reduced in size to dimensions below a quarter wavelength, so the radiation pattern tilts upward (as shown in Figure 7.8) and the radiation resistance drops.

It will be remembered that cables are usually 50-ohm impedance and this can be seen to be a compromise between the 73 ohms of a dipole and the 37-ohm quarter-wave. In practice, most antennas are designed for a minimum VSWR into a 50-ohm load. This means that the antenna will include an impedance-matching device and that for simple antennas (like whips) the elements will not be exactly electrical multiples of a quarter-wavelength long.

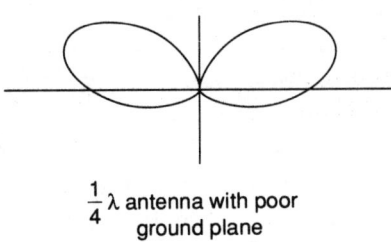

$\frac{1}{4}\lambda$ antenna with poor
ground plane

Figure 7.8 *The effect of the ground plane on the radiation pattern of a quarter-wave antenna.*

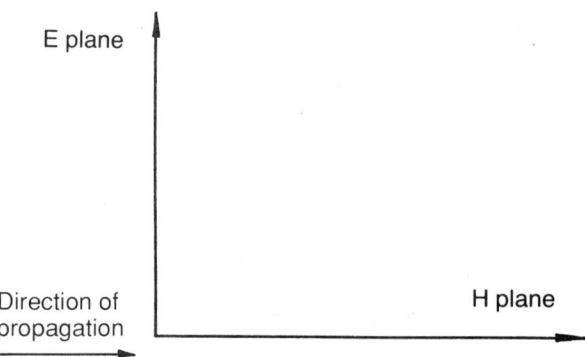

Figure 7.9 *Polarization of cellular systems.*

POLARIZATION

Radio propagation is characterized by its polarization, which refers to the relationship between the plane of the electric vector (the E-plane) and its angle with respect to the vertical. The H-plane or the magnetic plane is at right angles to the E-plane and in the direction of the maximum radiation as shown in Figure 7.9.

It has been traditionally assumed that for mobile services the E-plane is vertically polarized and antenna design has been based on this assumption. This assumption is based on the fact that both the receive and transmit antennas are vertically polarized. Recently, however some investigators have begun to consider the effects of multipath on the polarization, noting in particular that reflections can cause up to 180-degree rotations of the E-plane vector.

When a vertically polarized wave is incident on a perfect reflector, as can be seen in Figure 7.10, the component parallel to the reflector undergoes a 180-degree phase shift while the component that is normal undergoes no phase shift. Where the reflector has more complex characteristics (for instance, resistance) the reflected phase changes are more complex.

Refraction is a little more complex, as there will always be a boundary condition that will include a reflection. Continuity at that boundary will require that the vector sums of the incident and reflected waves equal the transmitted wave vector.

In short this means that virtually every reflection and refraction will result in a polarity rotation so that a mobile radio wave in the far field can be expected to have undergone significant E-plane twisting

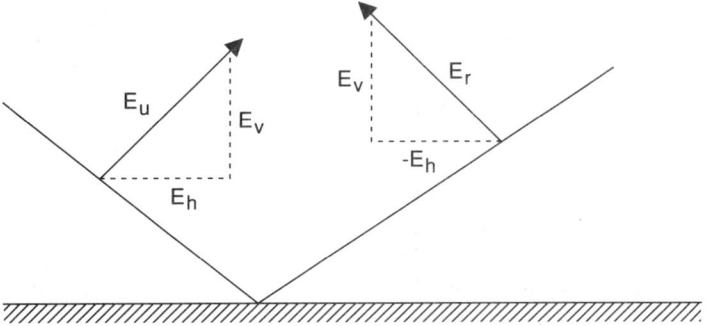

Figure 7.10 *The polarization of the E-plane after reflection from a perfect reflector.*

so that its polarity will be indeterminate. Some current studies are aimed at determining the nature of this twisting.

LEAKY CABLES

Leaky cables, which are effectively inefficient antennas, can be used to good effect in tunnels and buildings. In fact it is becoming a common practice to install leaky cables in new railway tunnels as a matter of course in anticipation of their future use by mobile services.

A leaky cable is one that is designed to radiate a portion of the signal carried along its length. Of course to some extent all cables are leaky since they will radiate some RF along their length. The main mode of radiation at high frequencies is due to magnetic flux leakage through the shield. Braided cables are far more leaky than solid cables as the "open" areas between the braids are the launching points for magnetic leakage. Leakage can be provided by cutting small slots in a coaxial cable as shown in Figure 7.11. The amount of loss can be controlled by the choice of slot size and regularity along the cable. This same technique can be used in high-frequency waveguide.

Cables also leak power by dissipation of the carried current, as heat due mainly to the conductor resistance and dielectric losses. This loss still has to be minimized as there is no RF radiation associated with this loss and as a consequence larger leaky cables will be more efficient than smaller ones.

When using these cables one should be aware that the leakage will not be uniform with distance—for example if a cable has a loss of 10 dB per 100 meters and is fed by a 100-watt PA, then 90 watts will be radiated in the first 100 meters, but only 9 watts in the next 100 meters and 0.9 watts in the next, and so on. Consequently, if long cable runs

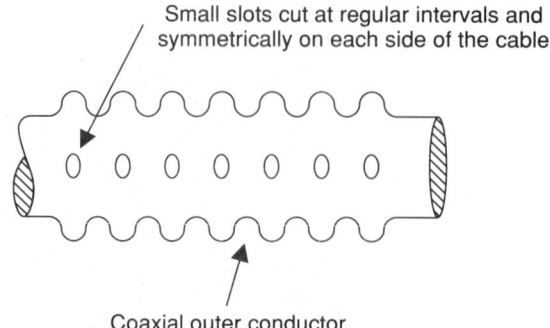

Figure 7.11   *A leaky cable is designed so that the shield has gaps to permit a regulated RF leakage.*

are planned it will be necessary to use not a single cable but a number of cables connected so that the first span is relatively low-loss, and with each successive span being slightly higher loss so that the average power leakage per meter is approximately maintained at a constant.

If repeaters are used one should keep in mind that they will be linear amplifiers and as such very prone to intermodulation distortion. To avoid this they need have the power backed off to the linear region (which is usually about 3 dB below the maximum gain).

Often it will be necessary to share a leaky cable with other users. There may for example be a trunked radio or other RF facility operating in the same region as the paging system. Where this is the case, some specialized repeater modules, like the one shown in Figure 7.12, that house repeaters and filters for two systems intended to work from the one leaky cable.

ANTENNA MATERIALS

Because antennas are exposed to the elements (sun, rain, ice, smog), the choice of materials is critical. All metals must be electrolytically compatible or else local corrosion cells will form. Joints with corrosion are potential intermodulation sites. Paging omnidirectional antennas usually have about 9-dBd gain and are of collinear construction, encapsulated in a fiberglass radome.

The radome must also be of high-grade fiberglass because water leakage can lead to corrosion of the elements and the ultimate failure of the antenna. Many fiberglass products, however, have metallic additives that make them unsuitable for antenna construction.

Figure 7.12   *A two-frequency leaky cable repeater designed to operate simultaneously at 450 MHz and 900 MHz. Photo courtesy of Andrews Australia.*

MOUNTING

Paging antennas must be properly mounted to function effectively. Omnidirectional antennas should be mounted on top of the tower or building as shown in Figure 7.13, or well offset from the tower, because side mounting can seriously distort the original omni pattern.

The connection between the main feeder and the antenna should include a tail of 2–3 meters, as shown in Figure 7.14. The tail makes installation easier and is important for future maintenance, when it may be essential to replace an antenna with one of a different type.

DRAINAGE

Drain holes must be clearly identified and correctly oriented. Because even the best-constructed antennas will suffer some water leakage, antennas are usually fitted with small drain holes. These holes are

Figure 7.13   *An omni-antenna, if it is to be side mounted should be offset from the tower as is the case for this omni. Caution should be exercised with this type of mounting because it is prone to movement in high winds and may need regular maintenance to keep the antenna vertical.*

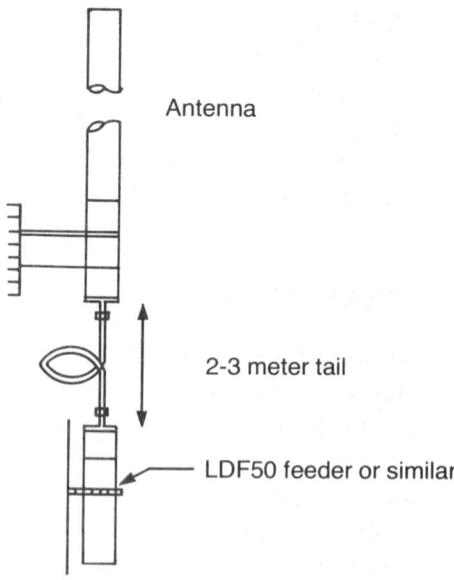

Figure 7.14   *The connection between the feeder and antenna should be by a 2–3 meter tail.*

located at the lowest points of the antennas. This is an important factor when considering mounting the antenna upside down (as is often done in cellular installations). If the antenna is not designed to be mounted this way, there will be a drain hole only in what was the bottom (and is now the top). In these cases, it is necessary to seal the existing drain hole and carefully drill a new hole in what was the top (and is now the bottom).

INTERMODULATION

When two signals of a different frequency mix in a non-linear device (for example, a rusty wire, loose joints on towers, and rusty fences), the result is intermodulation. Intermodulation can be a problem at any site that has two or more transmitters. Sites, which often have co-sited paging, trunked radio, and other mobile services, are likely to experience intermodulation. This problem may appear as interference to other (nearby) mobile service users.

Finding the intermodulation source can be very time-consuming and usually proceeds by eliminating likely offenders by trial and error. Because the problem is often intermittent, it can be very frustrating to track down.

Rusty bolts have long been blamed as the major culprits, but recent tests have revealed that the worst offenders are rusted galvanized mild-steel rope, mild-steel chains (the very worst offenders), and mild-steel wire fences.

Loose joints with small areas of contact are the main intermodulation points. Large areas of corrosion, such as on decking or galvanized iron sheets, may not necessarily produce high levels of intermodulation.

MEASURING VSWR

Antenna VSWRs should be in the range of 1 to 1.7, and each antenna should be checked upon installation. Most base stations have a built-in VSWR meter.

To measure VSWR, it is necessary to distinguish between forward signals and reflected signals. In a VSWR meter, a directional coupler is used to selectively read the signal in either direction. If two couplers are used, it will be possible to get a direct reading of VSWR by using some simple circuitry to derive the ratio of the forward and reflected signals. Figure 7.15 shows a simple directional coupler.

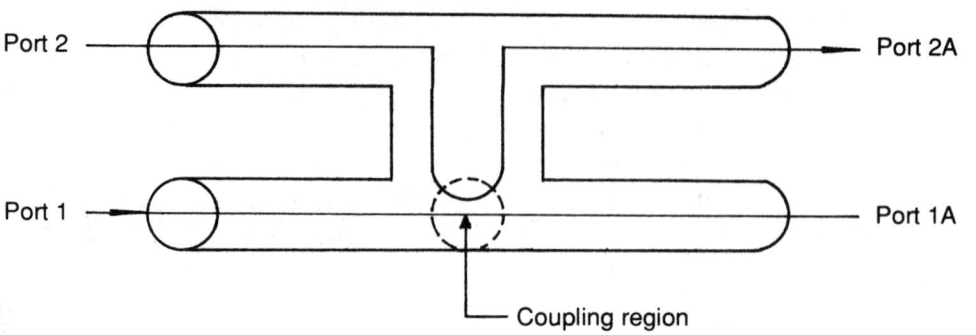

Figure 7.15   *A simple directional coupler.*

A directional coupler relies on placing the sensing loop so that the induced currents from the electric and magnetic fields are equal. The induced currents resulting from the electric field are indifferent to the direction of that field. The magnetic fields, however, are of opposite phase in the forward and reflected waves and will thus cancel the electric field in one direction. By specifying port 1 or port 2 as the sensor, a directional coupler can be made to read either the forward current only or the reflected current only.

 MULTIPLE TRANSMITTER OPERATION—WIDE-AREA COVERAGE

Paging operations in all but the smallest cities or areas will require the use of multiple transmitter sites to ensure coverage over wide areas. Paging receivers, because of their relatively inefficient antennas, require high field strengths for reliable operation, compared to other mobile receivers. This is partly compensated for by decoders that can work in a very poor signal-to-noise (S/N) environment, often achieving sensitivities of a few dB S/N.

The operational environment for pagers includes the interior of buildings, basements, and elevators. Because there is no confirmation procedure in paging, the user will never have acknowledgement of the message and so must trust that it succeeds. Because of this, a very high intrinsic reliability must be built into the RF design.

The two basic solutions are brute power, which involves using up to 3000 watts (depending on local regulations), and multiple sites. Brute power is often effective in rural environments and areas where there are no significant local obstructions. But where hills or man-made barriers exist the effective answer is usually multiple bases.

This is particularly true in high-density CBDs where penetration into the concrete jungle can only be assured by transmitters located in close proximity to the target area.

The usual way to operate a number of paging transmitters to give wide-area coverage is to use simulcast transmission. In this mode the transmitters in the network operate in parallel. The main problem is a tendency for the transmissions from overlapping sites to interfere destructively an so cause local dead spots. This interference can be caused either at the RF frequencies or at the modulation frequency.

The potential RF interference is relatively easily compensated for by arranging that the transmitters be offset in the center carrier frequency by around 600 Hz with respect to any neighboring bases and by 300 Hz for bases separated which are not neighboring, but nevertheless may be capable of interference. The modulation has to be time equalized using delay lines to ensure that the received signals are in phase to at least ±90 degrees.

DESIGN OF SIMULCAST SYNCHRONIZATION

The first task facing a designer of a simulcast system is to find the area where interference potentially may occur. This area is usually taken to be in the RF propagation path where the field strength of the overlapping transmitters are within 6 dB of each other. In reality the signals need to be within 3 dB to cause significant interference, but, remembering that the accuracy of even the best measurements is ± a few dB, considering the 6-dB area of overlap is prudent.

As seen in Figure 8.1, there may be a number of areas where interference may be a problem. The field-strength contour maps are plotted and the problem areas identified. At this point it is worth noting that many designers bypass this step and simply assume that the area where equalization is required is midway between the transmitter and its potential interferer. Although somewhat crude, the method is often effective in practice. It is particularly ineffective when local terrain is such that the propagation from the sites under consideration is not uniform, and so the assumption that the region of equal field strength is the midway point is not justified.

The interference is most easily understood in the instance where the signal is AM modulated. The principle is the same for FM and FSK but because it is more easily visualized in the AM case we will, for this exercise, assume AM modulation.

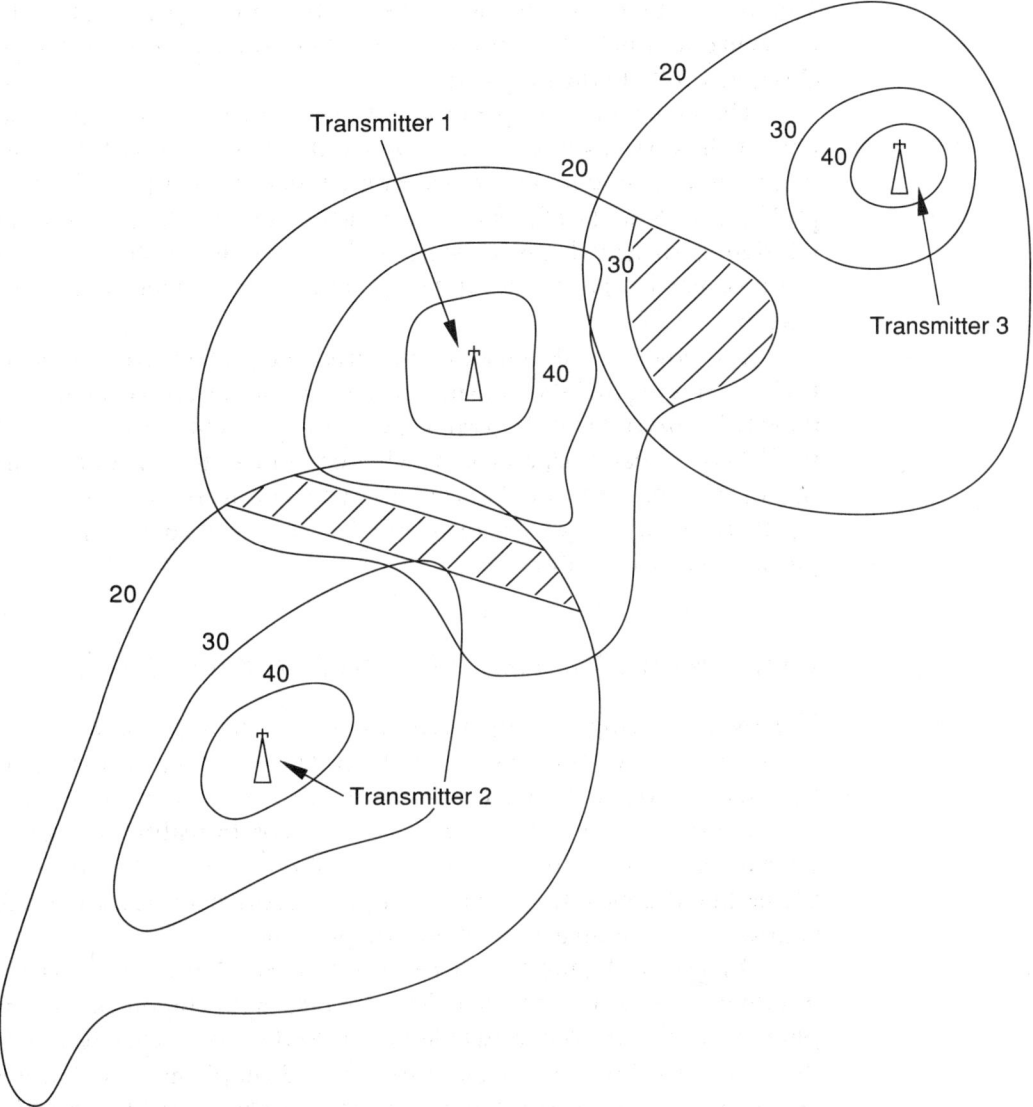

Figure 8.1

Consider Figure 8.2, where two simulcast transmitters are in phase, and the resultant signal seen by a receiver is the sum of the two signals. This is the desired effect in the areas of overlap. That is, the two signals reinforce each other. However, in the case where the signals are exactly out of phase, they will interfere destructively and cause local nulls (see Figure 8.3). One consequence is that a system that is properly synchronized can become destructive with a 180-degree phase reversal. FM systems are, in fact, very prone to failure

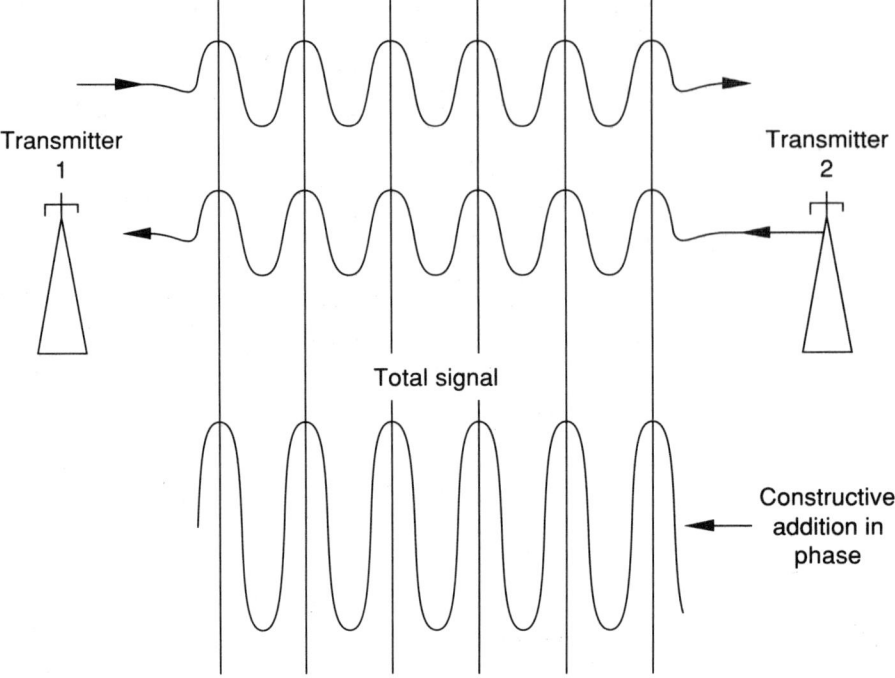

Total signal

Constructive addition in phase

Figure 8.2

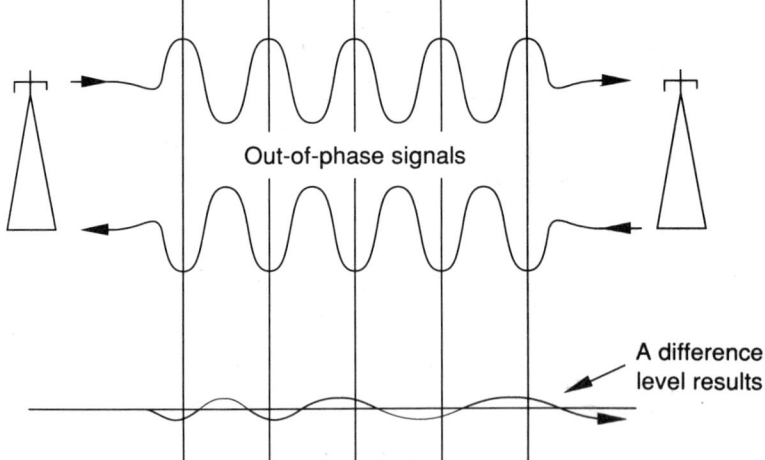

Out-of-phase signals

A difference level results

Figure 8.3

due to a simple reversal of audio phase at one of the transmitters, which transforms the situation shown in Figure 8.2 to that of Figure 8.3.

In order to achieve synchronization, delay lines are employed. These are usually located at the transmitter sites and are often built

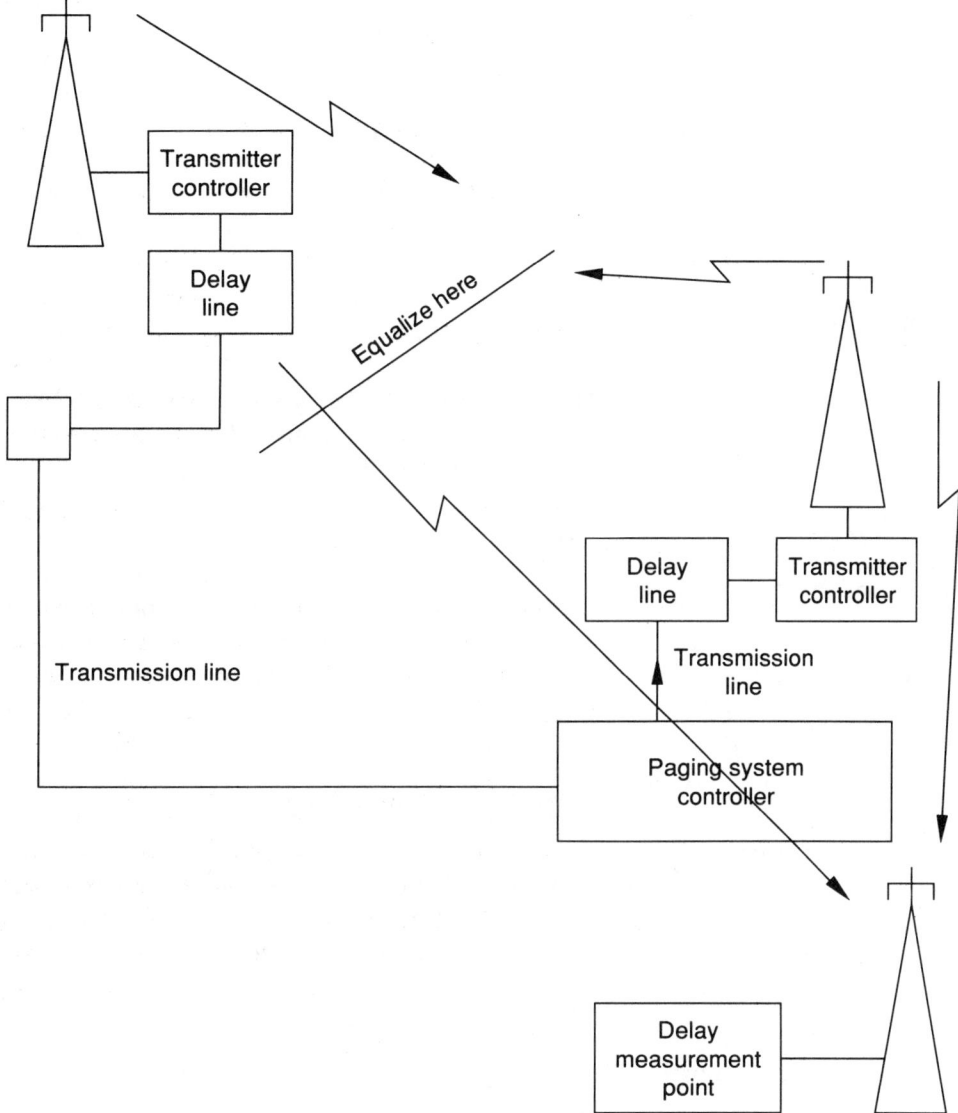

Figure 8.4

into the paging transmitter controller. From Figure 8.4 the components that contribute to the delay can be seen to be the transmission
line, the delay line, and the free space delay. Except for the transmission delay, these are easily quantified. The overall delay can be measured at a monitoring point (see Figure 8.4), and the net delay to the
equalization line can be calculated by subtracting the free-space propagation delay to the monitoring station and adding the free-path
propagation delay to the equalization point.

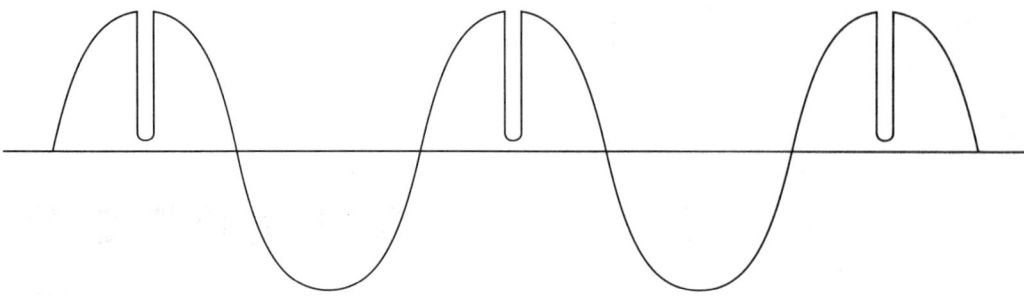

Figure 8.5

Using the fact that light and radio propagate at 300,000 km per second, it is easily calculated that the propagation delay in free space is *distance* (km) = delay (seconds)/300,000.

Measuring the Delay

It is not necessary to measure the absolute delay, as only the relative delay is important. One way to determine delay is to modulate the carrier with an audio tone of about 1 kHz. At this frequency, one cycle is equivalent to 300,000/1000 (300) km propagation in free space, which in turn is about 200 km propagation in a cable. This tone is generated with a marker to indicate phase. (Figure 8.5 shows such a signal.) If the monitoring point also is the point of generation of the modulation, then it is possible to adjust the delay lines until the transmitter being synchronized and the original signal are in exact phase. Once this is done for all the transmitters, it is only necessary to back off the delays to allow for the propagation delay time to the equalization point, and all the transmitters will be equalized.

Sequential Paging

Because simulcast paging requires precise frequency control, it is sometimes cheaper, in low pager-density areas (which nevertheless uses multiple transmitters for coverage) to use sequential transmission. When this method is used, it is necessary that each transmitter sequentially repeats all messages. For areas where there is overlap, this provides a crude form of time diversity.

For relatively small systems this method can be quite attractive, particularly as it can be converted readily to simulcast when the demand warrants.

CHAPTER

9

TRAFFIC ENGINEERING CONCEPTS

The measurement of telephone or circuit traffic and its application to circuit dimensioning (provisioning) is fundamental to all large-scale communications systems. Traffic is measured in Erlangs (most simply, an Erlang is one circuit in use for one hour). The traffic on one telephone line can be measured with an ammeter or voltmeter, as illustrated in Figure 9.1.

When a line is looped (that is, the handset is off the hook), a DC current of about 50 mA flows. This current can be detected by an

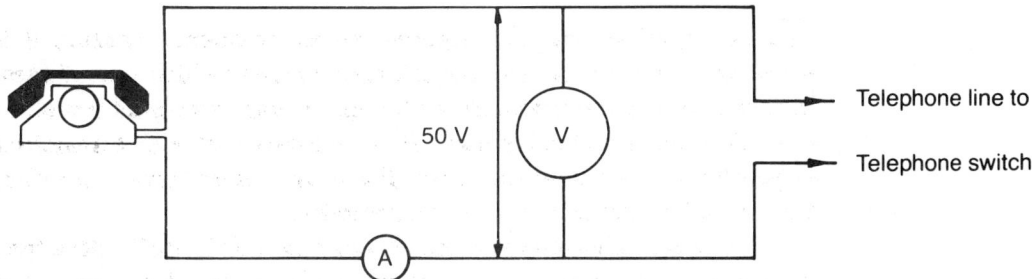

Figure 9.1 *A looped (in use) telephone drops the circuit voltage from 50 volts (open circuit condition) to about 5 volts.*

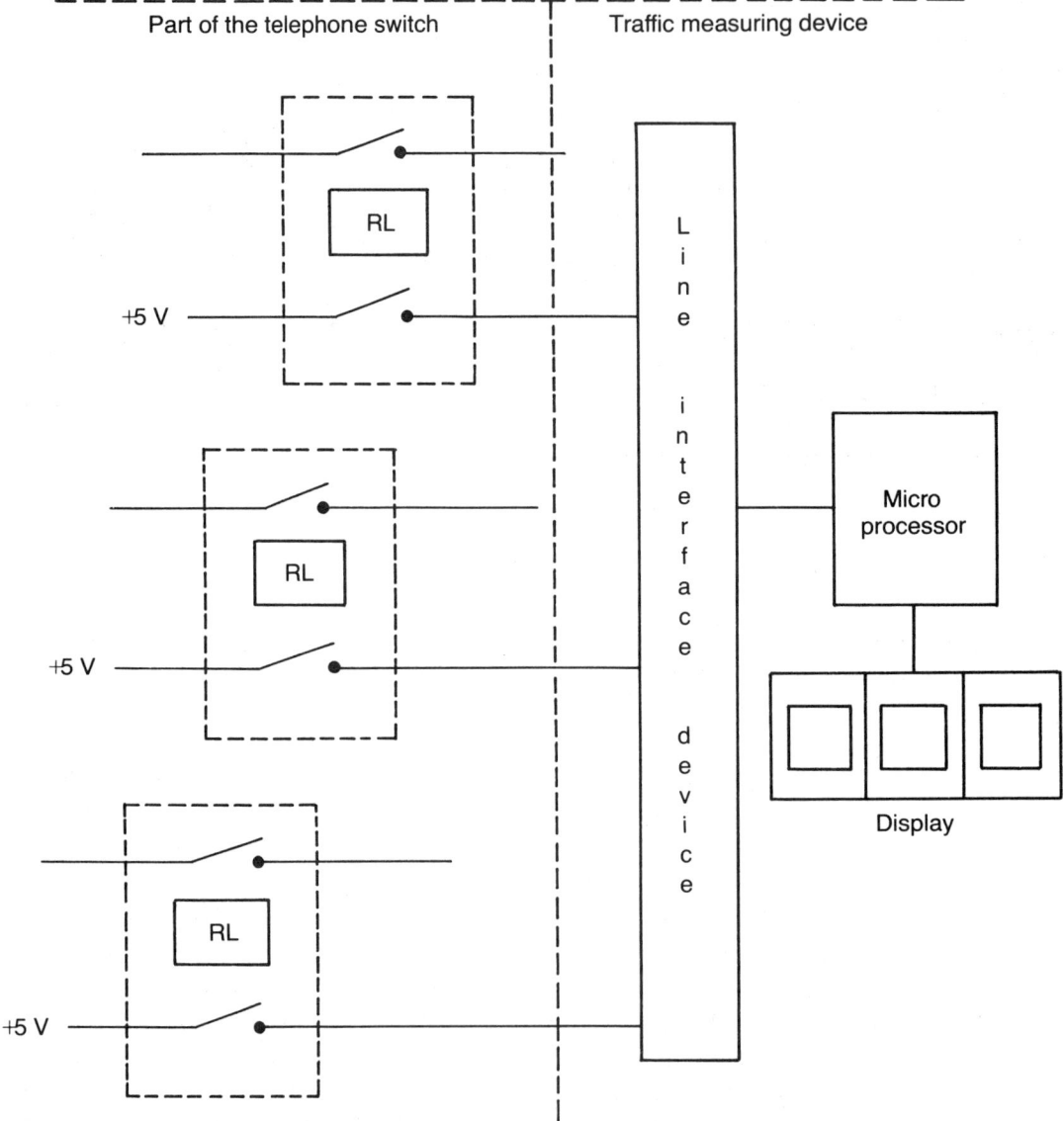

Figure 9.2   *The line relay RL, which activates when the telephone circuit is seized, can be used to measure traffic. A simple integrating device is used to measure the total traffic (in Erlangs).*

ammeter, and the DC voltage drops from the open circuit value of about 50 volts to about 5 volts. The actual loop current depends on the loop (line resistance), which can vary from 0 to 1500 Ω. When the handset is replaced, the current flow is zero and the line voltage returns to 50 volts.

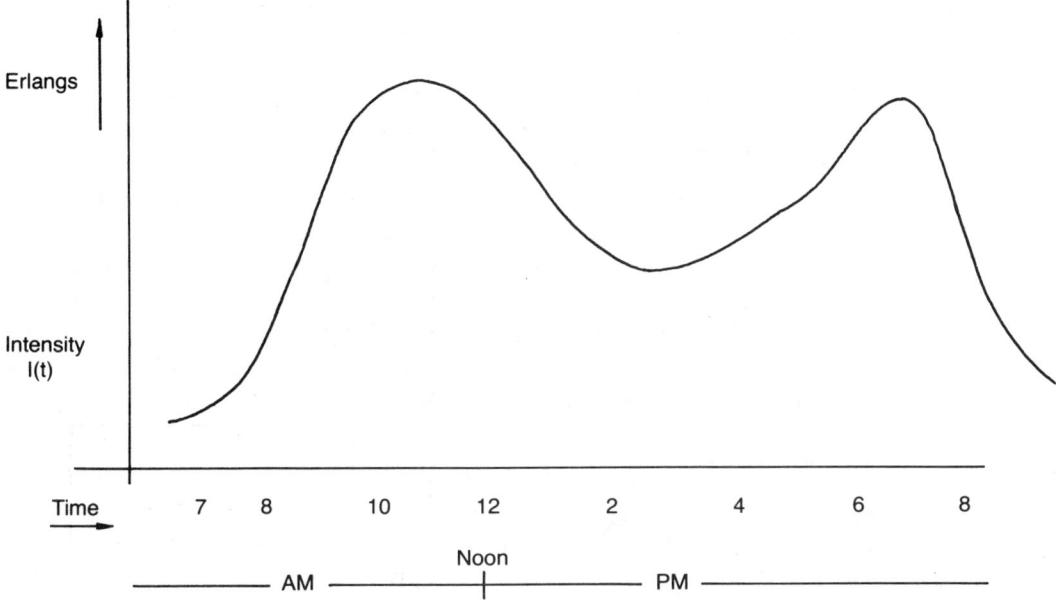

Figure 9.3   *The traffic density is shown here as a function of time of day. The traffic carried by a cellular system shows two characteristic peaks in the morning and late in the afternoon.*

A modern traffic meter uses a microprocessor connected to an interface that monitors many lines simultaneously, and it constantly scans each line to determine whether the line is in use. Figure 9.2 shows an example of a traffic measurement system.

Traffic measurements in paging systems should be examined on a monthly basis to ensure adequate network provisioning. Traffic can be represented by a number (of Erlangs), but remember that traffic varies with time and any practical representation is a compromise.

Fortunately, not much traffic engineering is involved in paging, but you should understand some basic concepts.

A telephone system will have a traffic distribution somewhat like the one shown in Figure 9.3. The total volume V of traffic carried, measured in Erlang hours is

$$V = \int I(t) \cdot dt$$

This integral is a crude measure of the cellular operator's income, particularly since the call charge is largely an air-time charge. Thus, V × air-time charge = air-time revenue.

This leads to the concept of call-holding time (the time that a call is held up). From the previous equation you can see that a call lasting one hour will generate the same traffic volume as 20 three-minute calls over the same period. The call-holding time of a typical cellular subscriber is 120–180 seconds.

The average calling rate per subscriber is the total traffic at the measured time divided by the number of subscribers. Ordinarily the rate is quoted in Erlangs per subscriber per hour. This figure determines the number of subscribers who can be placed on any given system, and it varies immensely. Average calling rates from 0.01–0.045 Erlangs/subscriber have been recorded and it would appear that the calling habits of different countries are very diverse. A calling rate of 0.025–0.035 could be regarded as average.

These rates are also quoted in milli-Erlangs (0.025 Erlangs = 25 mE). To calculate the total system traffic, it is merely necessary to multiply the average calling rate by the number of subscribers.

Notice that traffic varies in many ways with time. These are the most significant variants:

- Instantaneous variation of call arrivals.
- Hourly variations that depend on demand (usually early morning and late afternoon peak).
- A daily pattern is usually distinguishable. There is a marked difference between weekdays and weekends, and there may be significant day-to-day variations during weekdays.
- Seasonal peaks occur in the PSTN (for example, Christmas, Easter, Mother's Day).
- Tariff variations can cause significant (but usually temporary) variations in the call rate.
- Long-term variations in traffic can occur over periods of months or years.

Because it is not usually economic to provide circuits to cater to the peak demand, a compromise based on the provision of an acceptable grade of service in most instances is usually adopted.

Although traffic is measured in Erlangs to cater to the peak demand, in practice many different units of traffic are used. The two main ones are Instantaneous Erlang and Busy-Hour Traffic (in Erlangs). Instantaneous Erlang means the number of circuits in use at the instant in question. Busy-Hour Traffic means the average number

Table 9.1 *Busy-hour calculation*

TIME	TRAFFIC (ERLANGS)		
09.00	4.3		
09.30	4.8		
10.00	7.2) Largest adjacent readings – busy-hour traffic =		
10.30	6.1)	$(7.2 + 6.1)/2$	= 6.65 Erlangs
11.00	5.1		
11.30	6.2		

of circuits per unit time in use over the busiest hour. The traffic readings are usually taken in half-hour periods over the day, and the sum of the two adjacent half-hour blocks with the greatest total is defined to be the busy-hour traffic. This is why the busy hour usually starts either on the hour or the half hour. Table 9.1 shows the busy-hour calculation.

TIME-CONSISTENT BUSY-HOUR TRAFFIC (TCBH)

The concept of time-consistent busy-hour traffic (TCBH) is the one often used for dimensioning telephone circuits. It is only a concept and has no real physical meaning. Despite that, however, circuits are provided on this basis. It attempts to obtain a weekly average traffic measurement.

This concept is best illustrated by an example. Table 9.2 shows example base-station readings taken over a week. In this example, the traffic is measured as before, for every working day. Then, for each half-hour period over the five days, the average traffic measurement is found and put in the average traffic column. From the average traffic column, the busy-hour traffic is found as before; in this case, it is (5.3 + 5.6)/2 or 5.45 Erlangs. Notice that Monday's busy hour is the highest (6.65 Erlangs). The circuits are provided only for 5.45 Erlangs, so the nominated grade of service (probability of congestion) applies only to this theoretical TCBH traffic and not to any particular day.

Another definition of TCBH is to *define* the time at which the traffic is to be measured and measure all traffic at that hour. Typically peak-hour times such as 10 AM to 11 AM or 3 PM to 4 PM may be used.

Table 9.2   *Time-consistent busy-hour averages the busy hour over a one-week period.*

TIME	TRAFFIC MON	TRAFFIC TUE	TRAFFIC WED	TRAFFIC THU	TRAFFIC FRI	AVERAGE OF 5 DAYS
09.00	4.3	2.1	4.0	3.9	2.9	3.44
09.30	4.8	4.0	5.1	3.9	4.1	4.38
10.00	7.2) 6.65	6.1	3.0	4.7	4.7	5.14
10.30	6.1)	5.4	4.5	4.8	5.7	5.3)
11.00	5.1	6.4	6.1	4.8	5.6	5.6)*
11.30	6.2	3.9	3.1	4.8	4.0	4.4
* Highest adjacent sum equals TCBH traffic						

MEASUREMENT OF CONGESTED CIRCUITS

In congested circuits, it is likely that for significant periods of time all circuits will be busy. Because of the random nature of telephone traffic, even congested systems occasionally have free circuits and so the traffic carried is always less than the number of circuits.

Methods exist to determine the offered traffic on such circuits from the measured average traffic, but these techniques produce uncertainties that increase rapidly as the grade of service increases. Once the network becomes congested, it is difficult to measure traffic accurately enough to determine the offered traffic (and hence number of circuits needed) with any certainty.

DISPERSION

Dispersion measurements are used to indicate the sources and sinks (destinations) of traffic. Although direct occupancy measurements can give the total volume of traffic, they do not give any information about the direction or origin of that traffic unless all traffic on the route has only one sink.

Dispersion measurements involve analyzing the called number to typically six digits and recording the total holding time.

The switch will record the dispersion information on its statistics tape. From this a matrix of the originated traffic is obtained.

GRADE OF SERVICE (GOS)

A Grade of Service (GOS) figure is used to express the probability that a call will be lost due to switching or transmission congestion. Because it is a probability, the highest value it can have is 1.0; all calls will fail on any system that has this grade of service.

Like banks, which rely on the fact that it is unlikely that all their customers will require their money at the same time, telephone companies rely on the improbability that all their customers will attempt to place a call at one time. Banks will start a panic run if they are unable to meet the demand for funds at any time, so they must operate at a very low grade of service (that is, the probability of not having funds available to meet the demand at any time must be very low). For any one bank to do this alone would mean that it would need to keep a very large proportion of its funds available at any time in cash form. This would be very expensive. For this reason, vehicles such as the short-term money market exist to ensure that each bank has access to large sums at short notice without tying up too much of its own money.

Fortunately, telephone providers do not have quite the same problem. Usually, customers will just try again if their first call fails. But like the bank, these repeated attempts place a strain on the network, and too many repeated attempts can lead to total system collapse. The system must therefore be designed to minimize the possibility of such problems.

Telephone companies use a typical GOS range from 0.002 to 0.05. The acceptable range of call fail rates (due to equipment availability) is from 2 per 1000 to 5 per 100 attempts. For cellular purposes, the GOS varies from operator to operator, but 0.01–0.05 for base-station links and 0.002–0.001 for the switch to PSTN link are reasonable values.

DIMENSIONING SWITCH CIRCUITS

A GOS between 0.05 and 0.01 is ordinarily used. Table 9.3 can be used directly to dimension bases from measured traffic. A short BASIC computer program that can be used to calculate traffic for any GOS is found at the end of this chapter. This program is not copyrighted and can be freely used.

Table 9.3   *Erlang B table (continued next page)*

NUMBER OF VOICE CHANNELS	OFFERED TRAFFIC IN ERLANG FOR THE GOS SHOWN			
	0.001	0.002	0.01	0.05
1	0.001	0.002	0.010	0.05
2	0.05	0.07	0.153	0.38
3	0.19	0.25	0.46	0.90
4	0.44	0.53	0.87	1.52
5	0.76	0.90	1.36	2.22
6	1.15	1.33	1.91	2.96
7	1.58	1.80	2.50	3.74
8	2.05	2.31	3.13	4.54
9	2.56	2.85	3.78	5.37
10	3.09	3.43	4.46	6.22
11	3.65	4.02	5.16	7.08
12	4.23	4.64	5.88	7.95
13	4.83	5.27	6.61	8.83
14	5.45	5.92	7.35	9.73
15	6.08	6.58	8.11	10.63
16	6.72	7.26	8.87	11.54
17	7.38	7.95	9.65	12.46
18	8.05	8.64	10.44	13.38
19	8.72	9.35	11.23	14.31
20	9.41	10.07	12.03	15.25
21	10.11	10.79	12.84	16.19
22	10.81	11.53	13.65	17.13
23	11.52	12.27	14.47	18.08
24	12.24	13.01	15.29	19.03
25	12.97	13.76	16.12	19.99
26	13.70	14.52	17.0	20.9
27	14.44	15.28	17.8	21.9
28	15.18	16.05	18.6	22.9

Table 9.3 *Erlang B table (continued next page)*

NUMBER OF VOICE CHANNELS	OFFERED TRAFFIC IN ERLANG FOR THE GOS SHOWN			
	0.001	0.002	0.01	0.05
29	15.93	16.83	19.5	23.8
30	16.68	17.61	20.3	24.8
31	17.44	18.39	21.2	25.8
32	18.20	19.18	22.1	26.7
33	18.97	19.97	22.9	27.7
34	19.74	20.76	23.8	28.7
35	20.52	21.56	24.6	29.7
36	21.30	22.36	25.5	30.7
37	22.03	23.17	26.4	31.6
38	22.86	23.97	27.3	32.6
39	23.65	24.78	28.1	33.6
40	24.44	25.60	29.0	34.6
41	25.24	26.42	29.9	35.6
42	26.04	27.24	30.8	36.6
43	26.84	28.06	31.7	37.6
44	27.64	28.88	32.5	38.6
45	28.45	29.71	33.4	39.5
46	29.26	30.54	34.3	40.5
47	30.07	31.37	35.3	41.5
48	30.88	32.20	36.1	42.5
49	31.69	33.04	37.0	43.5
50	32.51	33.88	37.9	44.5
51	33.33	34.72	38.8	45.5
52	34.15	35.56	39.7	46.5
53	34.98	36.40	40.6	47.5
54	35.80	37.25	41.5	48.5
55	36.63	38.09	42.4	49.5
56	37.46	38.94	43.3	50.5

Table 9.3   *Erlang B table (continued next page)*

NUMBER OF VOICE CHANNELS	OFFERED TRAFFIC IN ERLANG FOR THE GOS SHOWN			
	0.001	0.002	0.01	0.05
57	38.29	39.79	44.2	51.5
58	39.12	40.64	45.1	52.4
59	39.96	41.50	46.0	53.4
60	40.79	42.35	46.9	54.4
61	41.63	43.21	47.9	55.4
62	42.47	44.07	48.8	56.4
63	43.31	44.93	49.7	57.4
64	44.16	45.79	50.6	58.4
65	45.00	46.65	51.5	59.4
66	45.84	47.51	52.4	60.4
67	46.69	48.38	53.3	61.4
68	47.54	49.24	54.3	62.5
69	48.39	50.11	55.2	63.5
70	49.24	50.98	56.1	64.5
71	50.09	51.85	57.0	65.5
72	50.94	52.72	58.0	66.6
73	51.80	53.59	58.9	67.6
74	52.65	54.46	59.8	68.7
75	53.51	55.34	60.7	69.7
76	54.37	56.21	61.7	70.7
77	55.23	57.09	62.6	71.7
78	56.09	57.96	63.5	72.7
79	56.95	58.84	64.4	73.8
80	57.81	59.72	65.4	74.8
81	58.67	60.60	66.3	75.8
82	59.54	61.48	67.2	76.8
83	60.40	62.36	68.1	77.9
84	61.27	63.24	69.1	78.9

Table 9.3 *Erlang B table (continued)*

NUMBER OF VOICE CHANNELS	OFFERED TRAFFIC IN ERLANG FOR THE GOS SHOWN			
	0.001	0.002	0.01	0.05
85	62.14	64.13	70.0	79.9
86	63.00	65.01	71.0	80.9
87	63.87	65.90	71.9	81.9
88	64.74	66.78	72.8	83.0
89	65.61	67.67	73.7	84.0
90	66.48	68.56	74.7	85.0
91	67.36	69.44	75.6	86.1
92	68.23	70.33	76.6	87.1
93	69.10	71.22	77.5	88.1
94	69.98	72.11	78.4	89.2
95	70.85	73.00	79.4	90.2
96	71.73	73.90	80.3	91.2
97	72.61	74.79	81.2	92.2
98	73.48	75.68	82.2	93.3
99	74.36	76.57	83.1	94.3
100	75.24	77.47	84.1	95.3

The path from the PSTN to the controller is usually dimensioned at either 0.002 or 0.001 GOS because any traffic lost on this path causes calls to fail.

CIRCUIT EFFICIENCY

It is important to note that when a small number of channels are used, the circuit efficiency (which can be defined as subscribers per circuit) is very low.

Converting traffic (in Erlangs) to subscribers is relatively straightforward. First, it is necessary to know the average calling rate of a subscriber. This is typically six calls/day of 150 seconds duration for a PSTN user, which translates to 0.03 Erlangs/subscriber (given

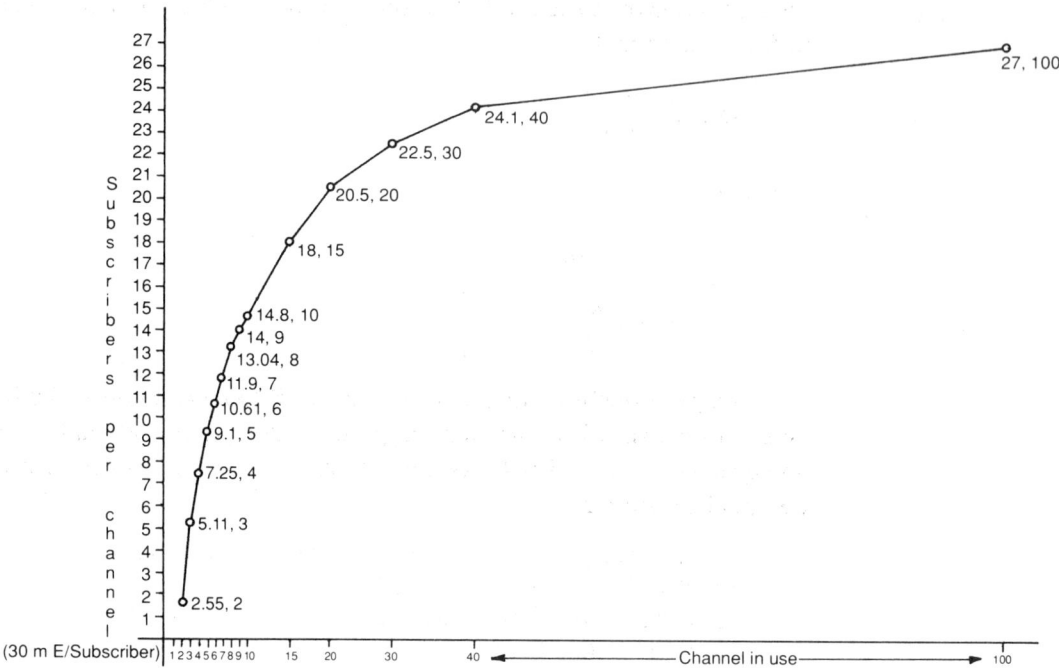

Figure 9.4   *The number of subscribers per channel increases rapidly as the number of channels in use increases.*

suitable assumptions about the busy-hour pattern). Hence the number of subscribers per base is

$$\frac{\text{Traffic capacity of the base}}{\text{Calling rate of subscriber}}$$

To show how the circuit efficiency varies with total circuits (channels), Figure 9.4 shows subscribers/channel versus number of channels for a 0.01 grade of service, assuming a calling rate of 0.03 Erlangs/subscriber (note that 0.03 Erlangs is often written as 30 mE).

ERLANG B TABLE

The Erlang loss formula is based on the probability of congestion B, being

$$B = \frac{P(N)}{P(0) + P(1) + P(2) + \ldots + P(N)}$$

where P(N) is the probability that the Nth circuit is busy when offered traffic A, such that

$$P(N) = \frac{A^N e^{-A}}{N!}$$

This simplifies to

$$B = \frac{\dfrac{A^N}{N_i}}{1 + A + \dfrac{A^2}{2} + \ldots + \dfrac{A^N}{N_i}}$$

As previously mentioned, the Erlang B table can conveniently be put into a simple computer program to allow ready calculation of any grade of service. The following BASIC program can be run on any personal computer.

```
10  REM program to produce Erlang B circuits
20  PRINT "Erlang B table"
30  INPUT "offered traffic" ; A
40  INPUT "GOS" ; G.
50  C = 1
60  N = 0
70  N = N + 1
80  C = 1 + N * C/A
90  B = 1/C
100 IF B > G THEN GOTO 70
110 F$ = "No. of CCTS."
120 PRINT USING; F$; N
130 GOTO 30
140 END
```

This program can be used to calculate base-station link and switch-junction circuits from known or estimated traffic and nominated GOS.

CHAPTER
10

SWITCHING

Until recently, it was relatively easy for a switching engineer to have little or no knowledge of radio systems and for a radio engineer to know nothing of switching. In paging, trunked radio, and cellular radio, however, the two technologies interact inseparably. The basic concept of a switch is to connect one line (usually a subscriber) to another in such a way that any subscriber (or line) can eventually connect to any other. When the number of connections is small, this can easily be done manually. Consider the situation of the four subscribers shown in Figure 10.1. By placing the link between any two subscribers (as shown between subscribers 1 and 2 in Figure 10.1) the operator can connect them in any order.

As the number of subscribers grows, the operator's task becomes increasingly difficult; automatic telephone exchanges are needed to cope with the number of potential links.

The first automatic telephone exchange was produced in 1892 at La Porte, Indiana. It was electromechanical—electric switches driven by electromagnets which by mechanical movement performed the switching function. Switching was accomplished by sending pulses (dialing) to indicate the number required. In this sense, these early switches were digital.

The early switches were actuated by dialing the pulses directly, which caused them to step on to the line required (hence the name

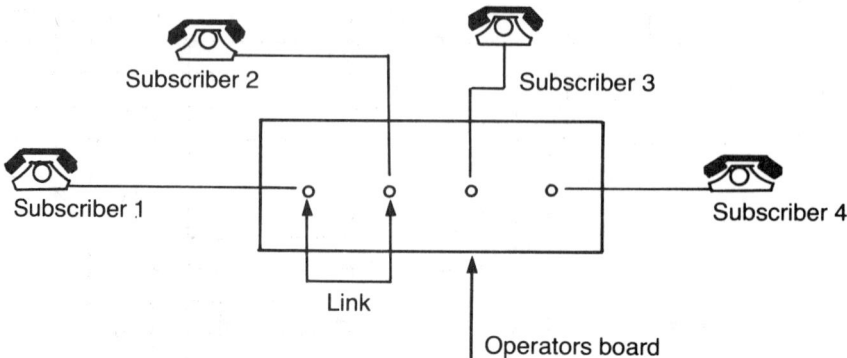

Figure 10.1   *With this simple operator-controlled switch, a manual connection between subscriber 1 and subscriber 2 is made by the link.*

step-by-step). Later systems used various forms of memory so that the exchange switching could be done asynchronously (at a different speed) to the dialing pulses. With refinements, this type of switching and dialing remained the standard until very recently.

SPACE SWITCHES

Telephone switches were originally all "space" switches; that is, the switches physically connected one circuit to another with a connection in space. In order to make a call between two telephones, it was necessary first to physically connect the two telephones with a wire. Space switches were used to connect subscribers for the duration of the call only, so that the links could be used by other subscribers after the call was completed. Today, although a physical connection usually does not occur (because the digital switches are time-multiplex-devices), two telephones are connected by a dedicated route to each other for the duration of a call.

Figure 10.2 shows a switch in which each inlet can connect to each outlet, as well as being able to park in a neutral position. This is known as a full-availability switch, since each inlet has a path to each outlet. The number of possible paths is 16 ($4 \times 4 = 16$).

As the switches get larger, the total number of possible paths rapidly increases. Consider a 400-inlet switch with 400 outlets; the total number of paths is 160,000 (400×400). Because of these huge numbers (and hence the massive amount of hardware), early switches were limited-availability, which means that each inlet could access

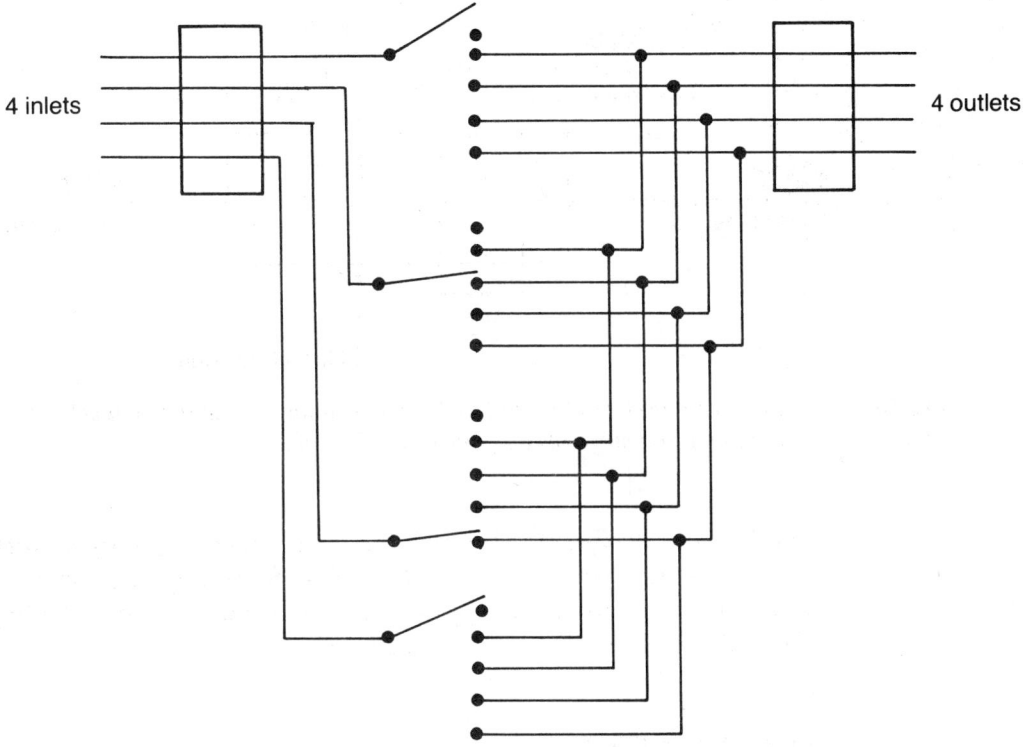

Figure 10.2   *A four-inlet, four-outlet switch. Inlets 1 and 4 are in the parked or neutral position, inlet 2 is connected to outlet 2, and inlet 3 is connected to inlet 4.*

only a limited number of outlets. For example, if the outlets per inlet are limited to 20, the total number of possible paths is 8000 ($400 \times 20 = 8000$), which is considerably more manageable.

TIME SWITCHES

Time switches became available with digital techniques. These switches work on the principle of switching a particular inlet to a particular outlet at a certain point in time. Figure 18.3 shows how inlets are assigned their respective timeslots. The input data is then rearranged (switched) under the direction of the control store so that each incoming timeslot is connected to the desired outgoing timeslot.

In Figure 10.3, each telephone line is sampled in its respective timeslot. The telephone in timeslot 1 on the A side is connected to the telephone in timeslot 4 on the B side. Notice that the switching is done

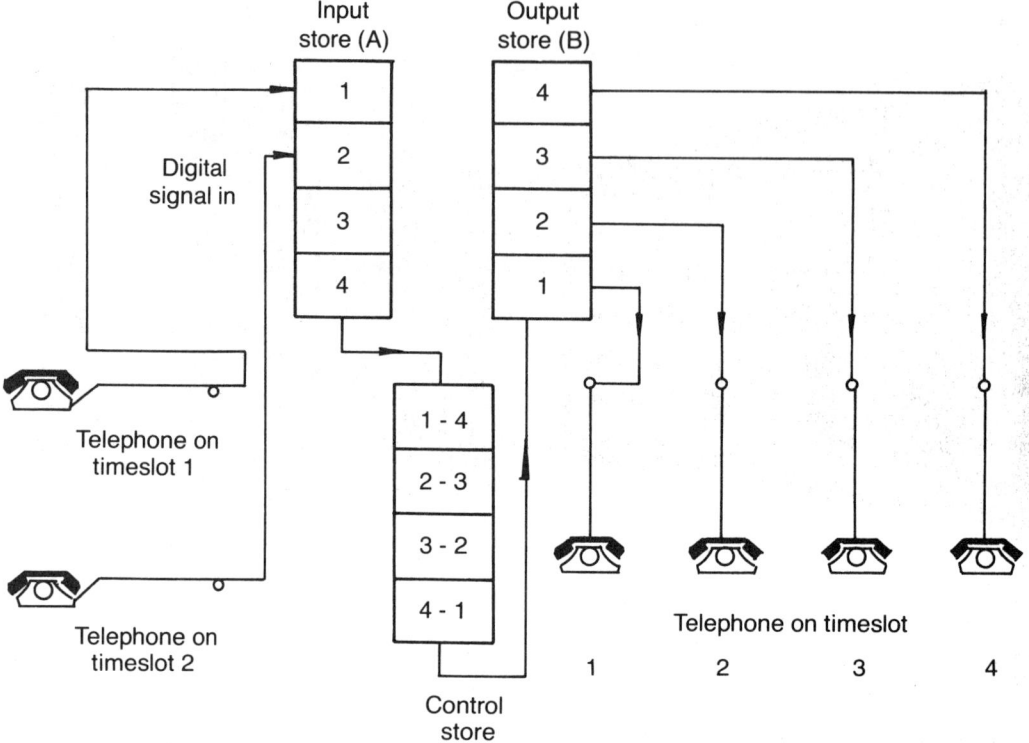

Figure 10.3   *A full availability time switch uses timeslots to connect any inlet to any outlet. The timeslot translation is shown in the control store. Inlet 1 is connected to outlet 4, inlet 2 to outlet 3, and so on.*

by rearranging the timeslots, not by physical wires, so that the information can be carried by a single path between switch A and switch B.

Modern telephone exchanges generally use a combination of time and space switches to minimize the total hardware needed.

SPC SWITCHES

SPC (Stored Program Control) switches ordinarily use both types of switching in the one switch. As Figure 10.4 shows, modern SPC switches come in a variety of sizes and are usually designed to be housed in air-conditioned rooms. The neat suites of equipment house the processor and the switch.

Figure 10.4   *A modern AXE10 from Ericsson—an SPC switch that can be used for conventional telephone or mobile telephone switching. Photo courtesy of L. M. Ericsson.*

SWITCH CONCENTRATORS

Figure 10.5 shows a simple concentrator switch. The switch is non-blocking, because every inlet has potential access to every outlet. Call blocking can still occur, however, because the number of simultaneous calls permitted is limited by the number of outgoing routes. In the switch shown in Figure 10.5, three simultaneous calls are allowed.

The eight inlets A–H are concentrated into the three outgoing routes (1–3) by activation of the cross-points; only one switch in any row or column can be closed at one time. This principle, called line concentration, is used extensively in all telephone switching. A radio base station acts as a line concentrator because it connects the mobile subscribers to the paging switch in such a way that approximately 50–150,000 paging subscribers can be connected to the switch by one channel.

The switches discussed previously have one very significant practical disadvantage: The connection between any inlet and any

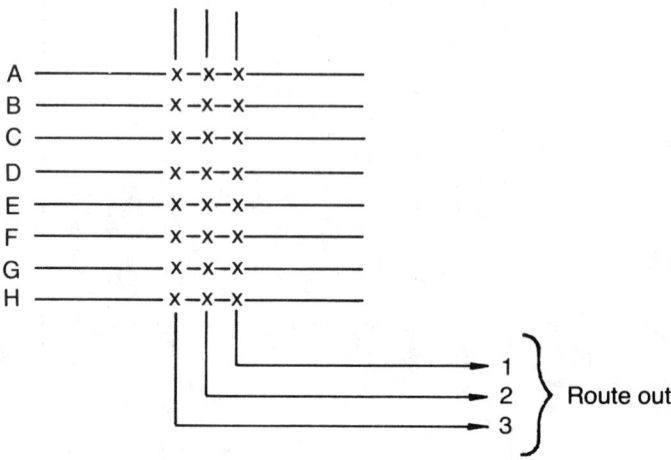

Figure 10.5   *It is easiest to think of a switch as a crossbar where connections are made by activating (connecting) the cross points. This limited-availability concentrator has eight inlets and three outlets.*

outlet occurs by only one path. The failure of any switch cross-point means that certain paths are no longer available. This limitation can be overcome, however, by introducing a second row of switches, as shown in Figure 10.6.

In Figure 10.6, you can see that the path between two ports (for example, B and D) can be connected by engaging the B and corresponding D row bass on any of the bars 1–3. For example B can be connected by engaging the cross connection B-1 and then 1-D or alternatively B-2 and then 2-D, and so on. This results in three paths connecting B to D. Although this configuration doubles the number of switches, it provides a very valuable redundancy in internal paths.

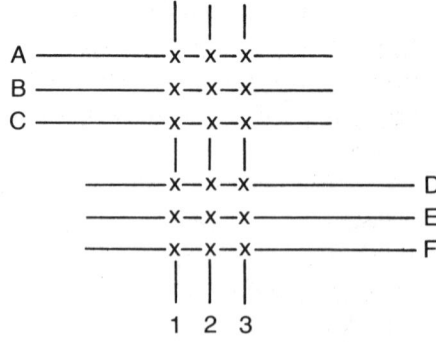

Figure 10.6   *This switch has multiple internal paths between inlets and outlets.*

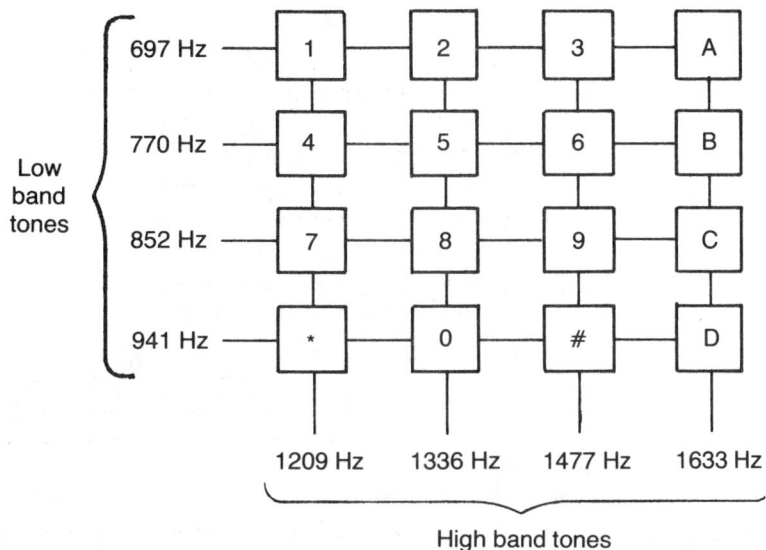

Figure 10.7 *DTMF, the standard for tone dialing, consists of two tones—one "low-band" and one "high-band"—generated from the matrix shown.*

DTMF DIALING (TONE DIALING)

Most telephones today employ DTMF (Dual-Tone Multifrequency) dialing. In this system, two tones are sent simultaneously to a line to indicate the desired number. Figure 10.7 shows the tone pairs and their associated numbers. The A, B, C, and D keys are not usually provided for POTs applications; they are reserved for special purposes.

DTMF dialing has been available for a number of decades. It was designed to take advantage of the potentially higher dialing speeds that were obtained by using code receivers with memory to store the digits and forward them as required by the switches.

In step-by-step (SXS) systems, each number dialed represents a train of pulses that cause mechanical switching in real time, making them necessarily slow. These systems are sometimes referred to as "stagger by stagger" systems by those who have seen the switches in operation.

With crossbar systems (and with modification, some SXS systems), code receivers were provided as exchange-common equipment. They were switched across the subscriber's line for the duration of dialing and could decode and ultimately store the DTMF pulses, which could then be sent further on in the network at any desired speed. Although DTMF dialing is a feature of almost every cellular telephone, it is by no means a new idea.

Each pair of tones in the DTMF scheme consists of one high-band tone and one low-band tone. This increases immunity to false decoding from voice or noise, as do other requirements such as a minimum signal-to-noise ratio and a correspondence in-level (known as twist) between tone levels for a successful decode. The tones are structured so that false decoding due to voice or noise is unlikely.

LIMITED-AVAILABILITY (BLOCKING) SWITCHES

Figure 10.8 shows the simplest limited-availability, or blocking, switch. In this example either light A or light B can be on, but not both. Because the switch can transmit information to only one of the outputs, it can lose information (for example, if a condition indicates that both A and B should be turned on simultaneously, some information is "lost" in the switch because the switch can only indicate the first state. A switch that can lose information is called a limited-availability or blocking switch.

The simplest telephone switch is an extension of the limited-availability switch, as shown in Figure 10.9. This simple uniselector switch allows a number of telephones to share a common outgoing line, but it has the disadvantage that only one telephone can use the line at any one time.

Subscribers' telephone switching stages will always be limited-availability switches. This is because one of the main functions of the subscriber's switch is to concentrate a large number of individual low-traffic telephone lines into a smaller number of high-usage lines that can be used to distribute the traffic efficiently. However, once the traffic is concentrated into parcels of about 0.5 Erlangs per circuit, it is

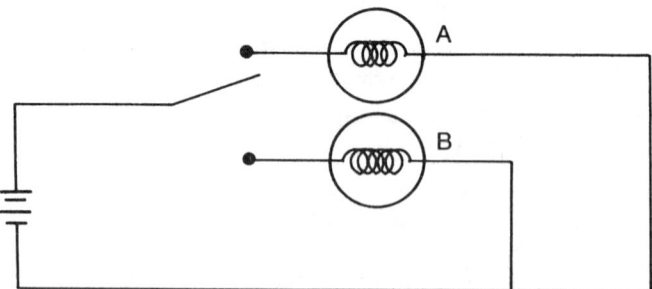

Figure 10.8   *Blocking occurs in limited-availability switches, and congestion or information loss can occur in the switch. The simplest blocking switch, shown here, enables one light or the other, but not both, to be turned on at the same time.*

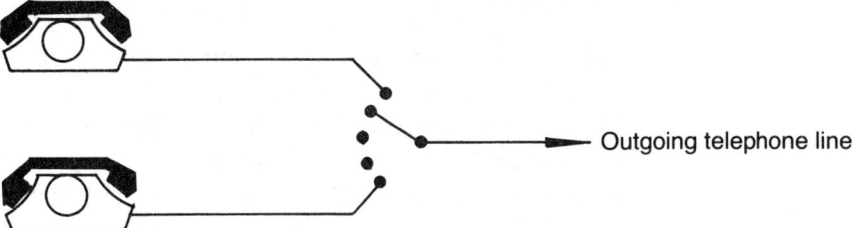

Figure 10.9   *A simple uniselector line switch (which is a line concentrator with limited-availability) can be used to concentrate telephones into a limited number of lines.*

efficient to use full-availability switches for onward trunking. Trunk switches are usually full-availability.

FULL-AVAILABILITY (NON-BLOCKING) SWITCHES

A full-availability or non-blocking switch is one through which it is possible to connect any idle outlet to any idle inlet, regardless of how many other connections have been made. Figure 10.10 shows the simplest full-availability switch. In this example, you can see that the switching of states A and B are independent and that information will not be lost through any limitation in the switch.

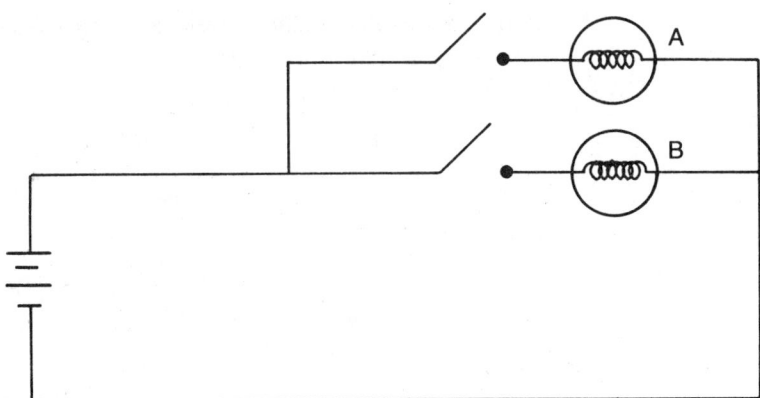

Figure 10.10   *The non-blocking switch shown here allows either or both lights to be in either the "on" or "off" state. No information loss occurs due to limitations within the switch.*

The switch concentrator shown earlier in Figure 10.5 is an example of a limited-availability or blocking switch. For example, if three of the inlets, A, B, and C, are connected to three outlets, 1, 2, and 3, respectively, then no other inlet can be connected until one of the established connections is dropped.

The switch shown earlier in Figure 10.6 is an example of a full-availability or non-blocking switch. The number of inlets and outlets must be equal in a full-availability switch.

A simple one-stage switch can easily and economically be made non-blocking for small-sized switches. Such a switch must have links from every inlet to every outlet, so the number of links increases as the square of the number of inlets. For large switches, this soon becomes prohibitive.

In 1953, Mr. C. Clos of Bell Laboratories published an analysis of three-stage switches, showing the relationship between the center switch configuration and the links. He demonstrated that for non-blocking it is necessary that each stage be non-blocking and that the number of center stages be:

Number of center switch points

$$= 2\,n - 1$$

$$= 2 \times (\text{the number of inlets/outlets per group}) - 1$$

Figure 10.11 shows a three-stage switch.

The path of any call may route from any inlet group to any center group by one link and from any center group to any outlet group by one link. Thus, there are K-paths through the switch from any inlet to any outlet.

It can be shown that the total number of cross-points for the switch system in Figure 10.11 is:

$$T = 2\,NK + K\left[\frac{N^2}{n}\right]^2$$

where

$N =$ the number of inlets/outlets

$n =$ the size of each inlet/outlet group

$K =$ the number of center arrays

The number of center arrays of switches can be determined by imagining a switch that has all circuits busy except one inlet and one outlet. In this instance, the worst-case is an inlet group which

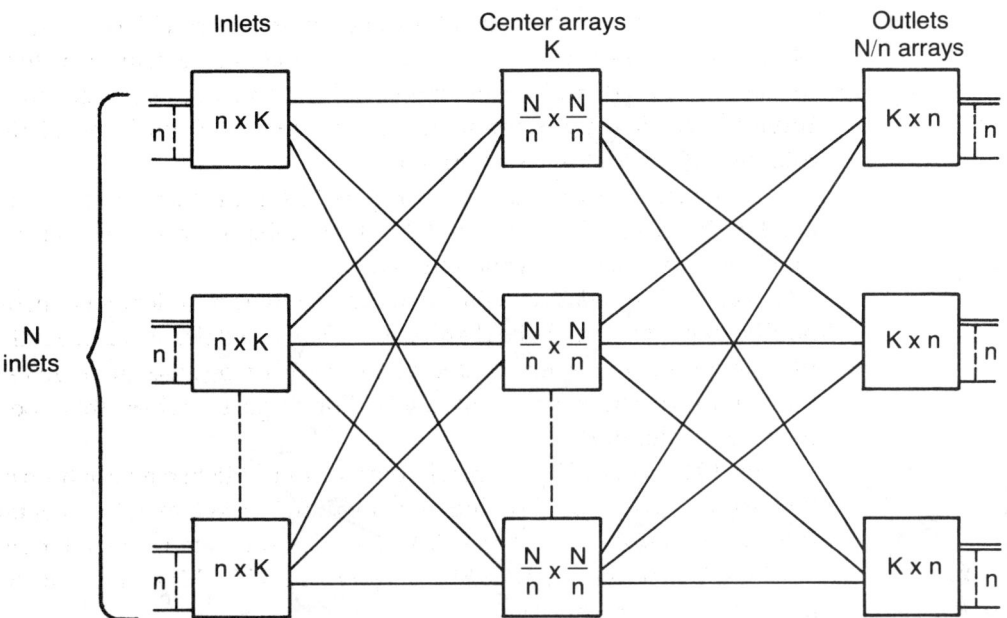

Figure 10.11 *This three-stage switch has K groups of n inlets and the same number of outlets. To ensure full-availability, there must be at least two (n − 1) switch cross points at the center.*

has n − 1 active outlets and attempts to connect to an outlet group that also has only one free outlet (the one sought), but that is accessed from a different group of center switches. Figure 10.12 illustrates this. To be full-availability, the switch must still be able to switch the path between the inlet and the desired outlet, so at least one other free path must exist.

The minimum number of center switch points is:

$$(n-1) + (n-1) + 1 = 2(n-1)$$

TRANSMISSION AND TRUNKING

Telephone traffic can be carried by one, two, four, or six-wire circuits. Each of these types of circuits has its place, and it is worth considering each separately.

Figure 10.13 shows a single-wire circuit. A one-wire circuit is an Earth-return system (the Earth provides the second wire). Early telephone links were mainly single wire, or SWER (Single Wire Earth Return), because this type of circuit was cheap. The disadvantages of

Inlets

Outlets

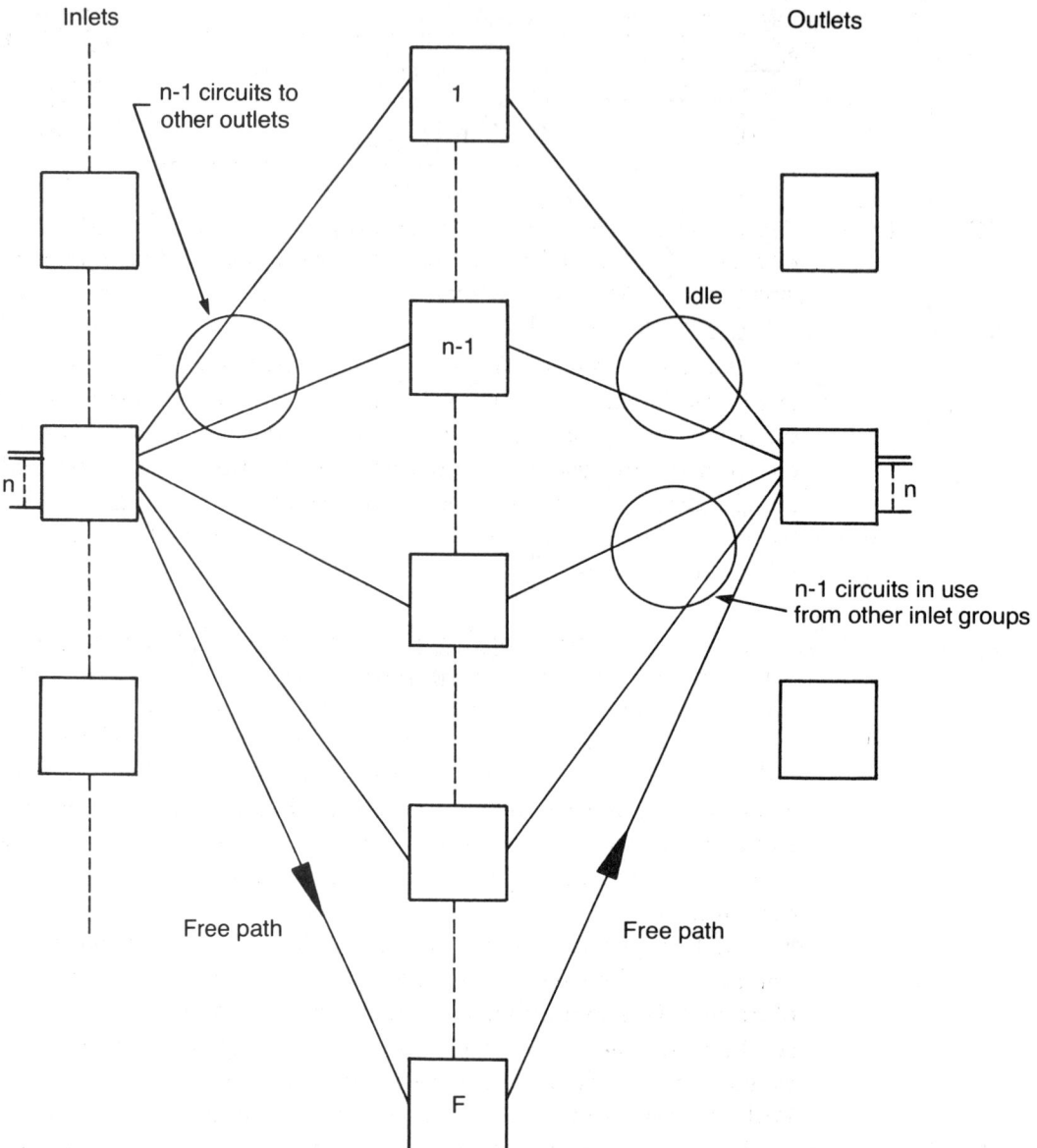

n-1 circuits to
other outlets

Idle

n-1 circuits in use
from other inlet groups

Free path

Free path

Figure 10.12   *This figure shows n inlets in one group that have n − 1 circuits busy to other outlet groups. The required outlet group also has n − 1 circuits occupied. To avoid blocking, at least one free path (via center switch F) must be available.*

such circuits are many and include variable performance due to soil resistivity, low noise immunity, and safety hazards in lightning-prone areas. Despite these problems, some single-wire circuits exist even today in rural areas.

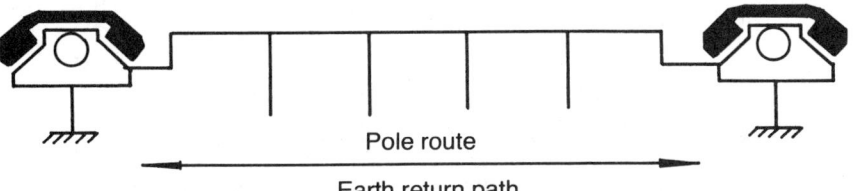

Figure 10.13 *The earliest (and cheapest) transmission systems used the earth as the return circuit. This method generated considerable noise and transmission-level uncertainties.*

Figure 10.14 shows a two-wire line. The two-wire line is an obvious improvement from the SWER line, and it does eliminate a lot of the SWER line problems. However, when longer routes are considered, a two-wire system has limitations when amplification is needed. Some ingenious amplifiers known as Negative-Impedance Repeaters (NIR) were developed to provide some amplification over long two-wire system routes. A conventional telephone is an example of a two-wire circuit.

The circuit shown in Figure 10.14 shows a subscriber telephone connected by a twisted pair to a telephone switch. Two-wire connections from subscribers' units to the telephone switch are usual.

This figure also shows an NIR between the two switches. The NIR provides some gain to compensate for line losses between the switches. The operation of an NIR is such that the maximum theoretical gain that can be provided on the route (end-to-end) is 0 dB (a gain of 1) before instability occurs. In practice, gains of –6 dB are more common.

On longer routes it is necessary to provide a considerable amount of gain to make up for system losses. The only practical way of doing this is to use four-wire transmission (that is, to separate the send-and-receive paths and amplify each). High-level (trunk-level) exchanges are usually four-wire switches, as shown in Figure 10.15. With four-wire switching, transmission without loss can be achieved in practice, with noise considerations and hybrid leakage limiting the total amount of gain achieved.

Figure 10.14 *A negative-impedance repeater produces gain in a two-wire cable and provides a moderately effective means of extending transmission distances in two-wire cables.*

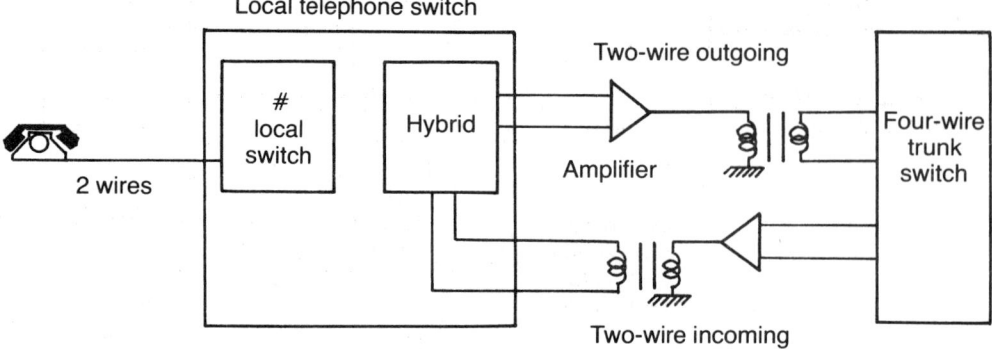

Figure 10.15   *The four-wire transmission (sometimes called six-wire or four-wire E&M) enables independent amplification of the send-and-receive directions. E&M are signaling media derived over the four-wire channels.*

Six-wire switching can be regarded as four-wire with two "wires" used for simple signaling purposes. (Four-wire links can derive the signaling channels in a number of other ways.) The two signaling wires are known as the M lead, which transmits the outgoing signal, and the E lead, which carries the incoming signal.

SWITCH CONFIGURATIONS

A telephone exchange consists of a switch to which 50 to 50,000 customers are typically connected. These subscribers normally have direct dialing access worldwide, which means the calls have a very wide dispersion. Consider a hypothetical town that has three subscribers' telephone switches, as shown in Figure 10.16.

If the town is relatively isolated, a fairly high percentage of the total traffic can be carried by the interexchange routes, so it would be justified to have direct trunks between those exchanges. When traffic to another area is considered (for example, to a town 200 km away), the traffic will be relatively light and the economics of providing three separate routes to the distant town may be rather poor. In this instance, a hierarchical switch, called a trunk switch, can be used effectively to concentrate the three traffic streams into one. Switch 3 in Figure 10.17 is a trunk switch.

All traffic routes from any switch in the town that are too small to justify a direct circuit can be switched through the trunk switch. In practice, the trunk switch may physically be one of the three subscribers' switches, but the trunk portion of it functions as the trunk switch shown in Figure 10.17.

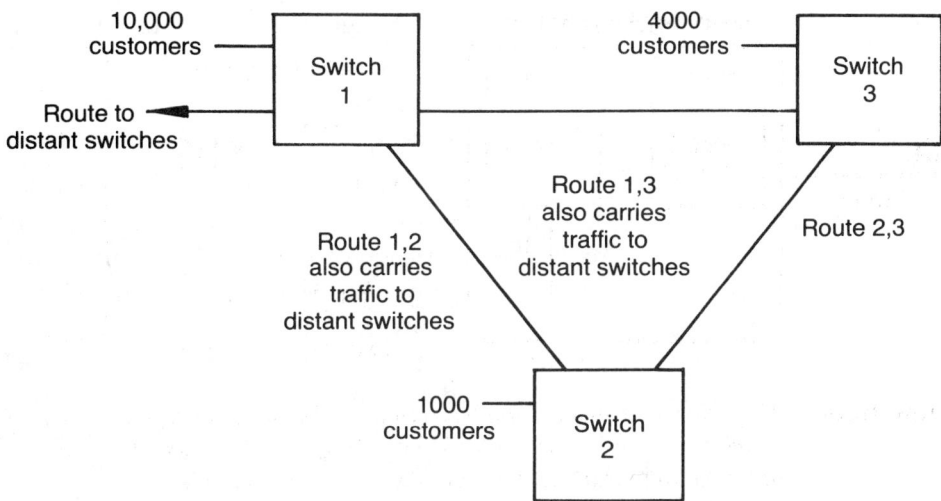

Figure 10.16 *Although it may be economical to directly connect local switches in a small town, it is often more economical to route all distant traffic via one (usually the largest) local switch.*

This principle is also applied to international calls; a few switches collect all international traffic and disperse it to distant destinations. At the distant end, calls are routed to their destination through successively lower-ranking trunk exchanges until they finally arrive at the desired terminal exchange.

Modern digital switches do not employ a rigid hierarchical structure, but rather are able to reconfigure their routing to take advantage of the best available route for any call. Thus, a local exchange can be both a terminal and a trunk exchange and can handle

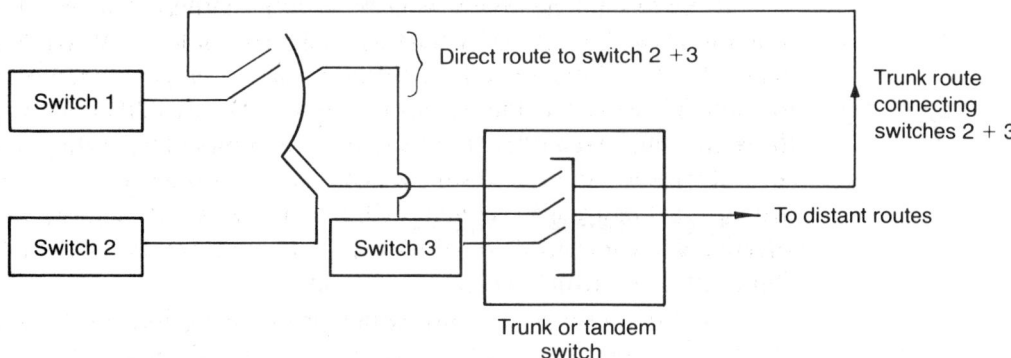

Figure 10.17 *A simple trunk switch configuration.*

transit traffic like a tandem. Exchanges that can perform this function are called nodes.

SWITCH HIERARCHY

It is recommended that the paging controller be placed high in the network hierarchy. The advantage of high-level trunking is that, in general, fewer switches are required for the average call, so trunking costs and transmission losses are minimized.

To visualize the relative position of a controller, consider a typical city network, as depicted in Figure 10.18. The controller ordinarily draws its customers from all over the city and therefore has no geographical center of interest (except perhaps the CBD). If the controller is connected low in the hierarchy—for example, at a primary center—then most calls must be routed through a number of switches before reaching their destinations.

As Figure 10.18 shows, the community of interest of a paging caller is the whole city, and most calls must be routed to a different area using the network hierarchy. This involves more switching paths and hence more loss than would be the case for higher-level switching.

Incoming and Outgoing Interfaces

Incoming and outgoing interfaces connect the switch to the outside world and provide the necessary signaling and signaling translation so the switch can communicate with other switches.

The signaling between the controller and the PSTN can take many forms. These are the forms most often encountered:

- Dial pulses
- MF
- MFC R2
- DTMF
- CCITT system number 7

To make matters even more complicated, the "standard" signaling systems can have many variations, and it is usual that the signaling at the paging controller must be tailored to the local PSTN version of the standard signaling format. This sometimes involves considerable software costs.

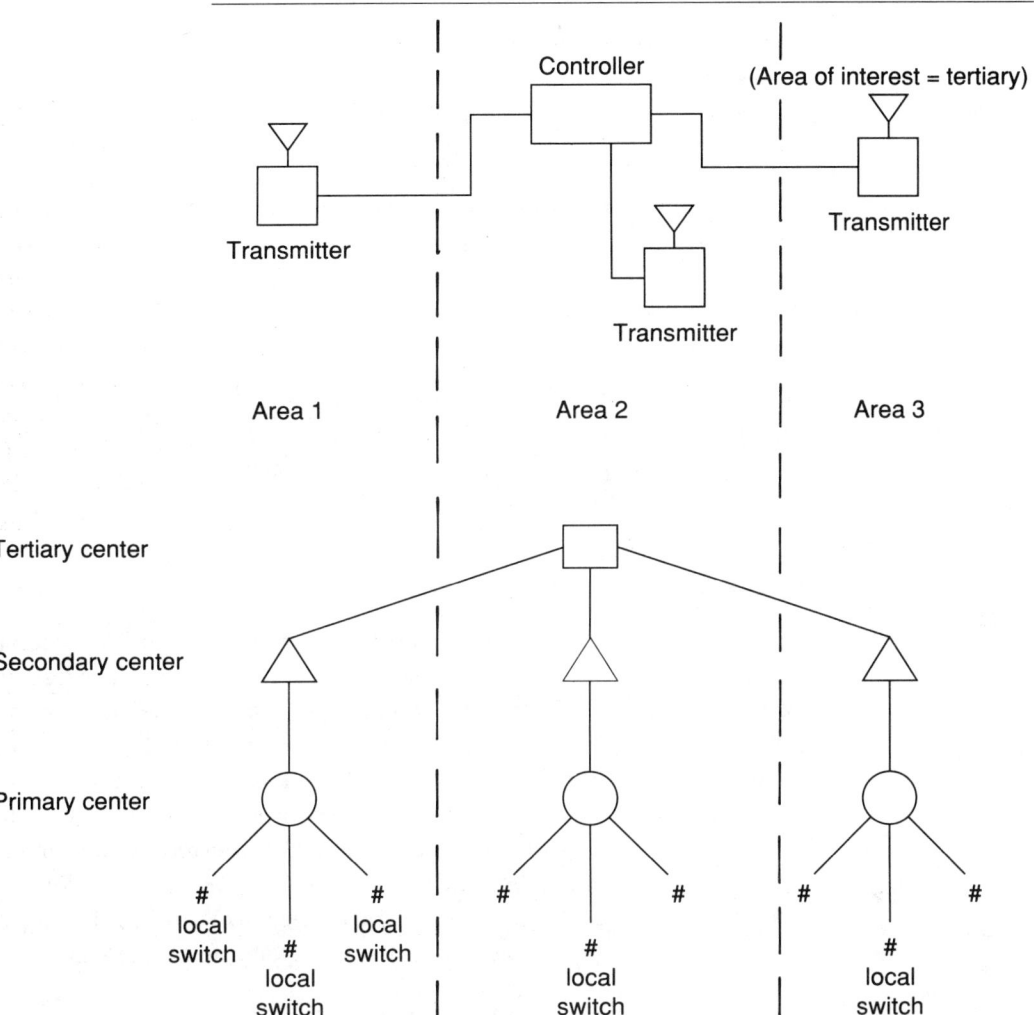

Figure 10.18   *The controller has a community of interest that includes the whole of the local service area of the PSTN. The controller should therefore connect at a high level in the trunk network.*

The controller will have limits on the usage of the outlets/inlets for their various functions. These functions provide for limited (but efficient) numbers of

- Both-way junctions
- Unidirectional junctions
- Total number of separate trunk groups
- Maximum trunks in a group

- RF, TX area groups
- Channels in each RF area group
- Intersystem data links
- Teleprinters
- Tape decks
- Voice recordings
- Tone receivers
- Operator positions

Any or all of these functions may have inherent limitations on the number of outlets available for each function; these limitations do not usually present problems, but they do restrict some fully loaded switch configurations.

Billing System

The billing system is a separate entity from the switch and may have no physical connection because the billing can be done by processing the raw data on the billing tape at a remote location. The billing computer for an average-sized operator is a minicomputer. If the operator is a wireline provider, however, the billing may well be integral with the wireline billing and use a mainframe computer. For real-time billing (that is, billing available instantly when requested), a data interface between the computer and the switch is necessary. Most billing systems will have limited real-time capacity so real-time billing is usually reserved for a subgroup of flagged customers. These customers are usually short-term renters.

The link may be RS232; like other such world "standards," RS232 has many non-standard forms, so the type of RS232 should be ascertained from the switch provider. The "standard RS232" is EIA-232D, published in January 1987, which conforms with CCITT V.24 and V.28 and ISO (International Organization for Standardization) IS 2110. These three references give functional, electrical, and physical standards, respectively.

The billing system may or may not also have an integral MIS (Management Information System) that also logs and analyzes system functions, such as channel usage, outages, traffic, and other housekeeping. The MIS may be interactive (that is, system commands such as subscriber validation can be input from the billing/MIS computer). Most billing systems expect a number of remote terminals to be operating simultaneously off the host computer.

Alarms

All alarms, both local and from the base stations, are reported to the controller. Often the controller features remote access to the alarm status to allow remote monitoring. Some systems also have remote access to each base station.

Alarms are usually divided into two or more categories, including major alarms (which affect service to a degree likely to be noticed by the user) and minor alarms (which may lead to partial reduction in capacity). The alarms are classified by the operation depending on the severity of the disruption to service of a particular fault. In a large city, the complete loss of one base station could be a minor problem, whereas in a small city that has only one base station the same loss would be a major problem. Lower levels of severity exist when, for example, a redundant unit is faulty and is switched out of service, or a simple channel is blocked in a large base station.

CONTROLLER LOCATION

Wireline operators have traditionally placed their main switches at what is known as the "copper center." The cheapest location for a land-line switch is the one that minimizes the total length of cable to the subscribers (and other exchanges). The location that minimizes the amount of copper (or total cable length) is known as the copper center.

Using a strictly mathematically accurate configuration to minimize the total length of copper is complex. However, the copper-center concept is relatively simple, as you can see in Figure 10.19. Here, the switch is located at the "center of gravity" of the copper mass feeding A, B, and C. For this reason, conventional wireline switches are usually found at the population centers.

It should be noted that paging controllers do not have subscriber links except for incoming calls. They also have links to the bases and the PSTN. Because only a relatively few links are involved—and with modern technology the cost of these links is not strongly distance-dependent—the concept of "copper center" is no longer dominant in choosing a switch site.

A wireline operator almost invariably chooses to place the controller in an existing high-security building, which for the wireline operator is the cheapest location. Such a site will be even more economical if the chosen building has a trunk switch (that is, the trunk link costs will be low).

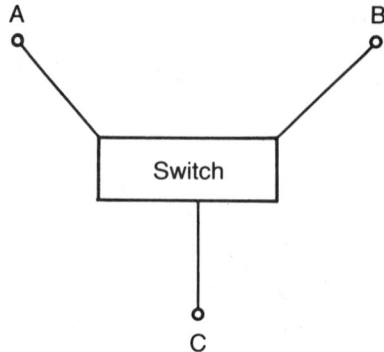

Figure 10.19   *The copper-center concept. A PSTN subscriber switch is placed at the copper center to minimize the total length of cable used.*

Also consider that the links to the base stations and the switch site should be well-located with respect to the bases. Base-station links are generally microwave or optical fiber. When microwave links are used, ease of link access to the switch site is a primary consideration. Whether microwave or fiber-optic links are used, the link cost is not strongly dependent on the link length.

NON-WIRELINE CONTROLLER LOCATIONS

For non-wireline operators, there is a wide choice of controller locations. Rental properties can be used for control rooms but are not recommended for the following reasons:

- Insecurity of tenure
- Lack of security where other tenants are involved
- Cost of moving the switch controller should it ever be necessary
- Unscrupulous landlords, who know the cost of relocation, could demand unreasonable rents
- Lack of control over the building, its expansion, and use

Purchased properties are much more practical for control rooms. Because there is no real need for a central location, costs can be controlled by locating the switch outside the expensive inner city area. A site that minimizes the total number of links will probably also minimize the cost.

SIGNALING

Signaling is the means by which information about the dialed digits, and the line condition and other network information is passed around the network. The early signaling systems were based on DC pulses; around 1950 a series of systems based on inband and out-of-band tones were developed. These tone-based signaling systems form the basis of the modern signaling formats although they now are more widely used in digital versions.

The most common form of signaling over trunks today is common-channel signaling, although voice-channel signaling is still widely used. In common-channel signaling, a different circuit is used for interswitch communications from that used for speech. Common-channel signaling enables switching to occur at very high speeds and means that voice channels are not tied up with signaling. The standard 2-Mbit, 32-channel system uses two of those channels exclusively for signaling, while a T1 uses speech channels with derived control channels.

In-band and out-of-band signaling use the same channels as those that are used for voice. Voice frequencies are assumed to be those in the bandwidth 300 to 3400 Hz. A commonly used in-band tone is the CCITT R1 signaling system, which uses a 2600-Hz continuous signal tone (with a notch filter to line for voice).

Although there are a number of standardized signaling formats it must be appreciated that within the standards there are user-definable spare codes. Almost every operator worldwide has managed to find some use for these spare codes and the interconnecting operator needs to carefully examine the local version of the signaling system, since often the absence of the optional code bits will cause signaling alarm errors. The paging software should be able to be set to accommodate any of the usual variants.

DC or Loop Signaling

Necessarily, the early signaling systems were very simple and were based on line pulses that alternately placed a "1" or "0" state on the line. In its simplest form, as seen in a rotary dial telephone, this is achieved by simply placing a short-circuit or open-circuit condition on the line. In the early step-by-step switches these pulses were translated directly into the physical movement of switch wiper blades, which in turn switched the call.

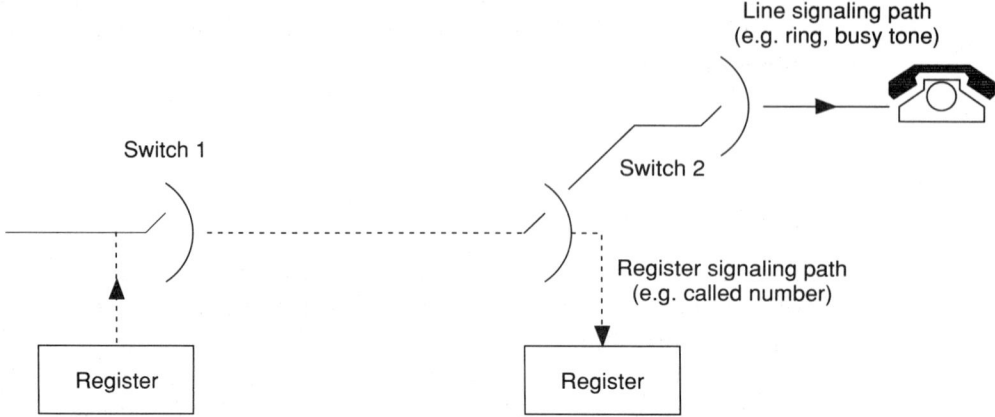

Figure 10.20   *The line- and register-signaling paths.*

Use is made of pauses to indicate the end of a signaling sequence. For example the pause between dialing one digit and the next on a rotary telephone is used by the system to determine the end of one string of pulses and the beginning of a new one.

Line-polarity reversal is used to indicate certain states including that the called party has hung up.

A major disadvantage of loop signaling is that the pulse waveforms deteriorate badly over long lines and that line pulses had to be separately reconstituted where amplification was used. This was usually done by converting the pulses to tones, which were ultimately converted back to pulses.

R1 Signaling

This form of signaling is composed of a line-signaling part and a register-signaling part. This is illustrated in Figure 10.20.

Line signaling is accomplished on the speech path either during the presence or absence of speech. The line signaling is done using a single 2600-Hz tone and its presence or absence is the basis for the signaling. The line contains filters at the subscriber's end to filter out the tone.

The registers form the "memory" of the switch and are used to store and forward the dialed digits. When a subscriber dials a number it is stored in the register, which can begin to switch the number as soon as sufficient digits have been dialed to determine which is the next switch.

Register signaling is done using two simultaneous tones chosen from a total of six. This is called *multifrequency code* (MFC) or sometimes known as "two out of six MFC."

This signaling is fully specified in the *CCITT Blue Book Fascicle vi.4*.

R2 Signaling

R2 uses an out-of-band signal of 3825 Hz for line signaling and a total of 15 forward and 15 backward inter-register signals. It is frequently used for both national and international signaling.

R2, like most other signaling systems, comes in both an analog and a digital version. The analog version is listed in Table 10.1.

Table 10.1 *R2 signaling tones*

		FREQUENCIES Hz					
SIGNAL	**FORWARD**	**1380**	**1500**	**1620**	**1740**	**1860**	**1980**
	BACKWARD	**1140**	**1020**	**900**	**780**	**660**	**540**
	1	x	x				
	2	x		x			
	3		x	x			
	4	x			x		
	5		x		x		
	6			x	x		
	7	x				x	
	8		x			x	
	9			x		x	
	10				x	x	
	11	x					x
	12		x				x
	13			x			x
	14				x		x
	15					x	x

Table 10.2   *R2 signaling Group A*

SIGNAL NUMBER	GROUP I SIGNAL	GROUP II SIGNAL
1	DIGIT 1	Subscriber without priority
2	DIGIT 2	Subscriber without priority
3	DIGIT 3	Maintenance equipment
4	DIGIT 4	Spare
5	DIGIT 5	Operator
6	DIGIT 6	Data transmission
7	DIGIT 7	International use
8	DIGIT 8	International use
9	DIGIT 9	International use
10	DIGIT 0	International use
11	Spare	Spare
12	Request not accepted	Spare
13	Access to test equipment	Spare
14	Spare	Spare
15	End of sending digits	Spare

The signal type can have two meanings, in the forward direction as defined under group I (selection) and group II (service class of calling party). Table 10.2 lists these meanings.

The backward signal direction also has two signal groups called group A (certification and control) and group B (status of calling party). These are detailed in Table 10.3.

It is fully described in the *CCITT Blue Book Fascicle vi.4.*

CCITT No. 5

This signaling system is meant for interswitch signaling and has largely been superseded by CCITT No. 7. It consists of one or two inband line signals and two out of six MFC inter-register signals.

A full description can be found in *CCITT Blue Book Fascicle vi.2.*

Table 10.3   *R2 signaling Group B*

SIGNAL NUMBER	MEANING OF GROUP A	MEANING OF GROUP B
1	Send next digit	Subscribers line free
2	Send digit before last	Subscriber has changed number
3	Change to reception of group B signals	Subscriber line busy
4	B signals congested	Trunk congestion
5	Send calling party's category	Unallocated number
6	Set up speech path	Metering line
7	Send second last digit	Non-metering line
8	Send third last digit	Called line out of order
9	Resend whole number	Spare
10	Spare	Spare

CCITT No. 7[1]

This is the most widely recommended interswitch signaling system, and it uses common-channel signaling (that is the signaling is done on dedicated channels).

For economical operation of a common-channel signaling system, signaling transfer points (STPs) can be used to provide dynamic routing of the signaling traffic.

No. 7 signaling is the most recent of the signaling systems and has been designed to allow interactive networking of switches. The information exchange using this format can be quite complex and interacting switches can be made to operate in a cooperative way that is not possible with the earlier signaling formats.

The digital form is a full-duplex signaling system operating at 64 Kbits/sec (56 Kbits in the SS7 version), an analog version that can operate over 3-Kbits channels is available.

No. 7 operates only on dedicated lines and other line conditioning such as echo suppressors and A or mu Law equipment must not be used.

---

[1] Known as SS7 in the U.S.A.

When performing high-level control it is similar to HLDC (high-level data-link control). Full details can be found in *CCITT Blue Book Fascicle vi.7, vi.8,* and *vi.9.*

INTERFACING SWITCHES

The signaling used between two switches will usually follow one or more of the above standards. For two switches to be interconnected, they must have a compatible signaling system.

In modern switches it is usual that all of the above signaling types would be available, but apart from E&M most of the others will be options that will cost around $40,000 to $100,000 (USD).

When purchasing a switch, it is essential to ensure that it will be compatible with the PSTN to which it will connect.

Devices do exist that can interconnect two dissimilar signaling systems, and these can be purchased as stand-alone devices.

It should be noted that the optional parts of the various signaling systems are widely used for custom purposes, and it will usually be necessary for the interconnecting operator to match the "modified" version used by the local PSTN.

SYNCHRONIZATION

Virtually all modern digital systems require synchronization in order to interwork. The paging controller and the digital transmission systems will need to be synchronized to the networks to which they interconnect.

The paging system can be either slaved to the PSTN, which presumably has a high-stability reference clock such as a cesium beam, or it may be synchronized to its own high-stability clock. If it is slaved, then it will be connected to two separate clock sources to ensure redundancy.

Large network operators will usually have a number (2–20 depending on the network size) of high-stability clocks distributed throughout the network to which the rest of the network can slave.

The slave system will usually use the master clock to synchronize its internal clock so that should the source of the master fail (as for example during transmission fades), its internal reference can keep the system running.

Where the clocks fail to synchronize there will be data loss as a frame slip, which can be either chronic, a Prolonged Frequency Offset

(PFO), or a Transient Synchronization Loss (TSL). The latter is more likely to occur when the slave clock overreacts to a noise pulse or fails to stabilize when a changeover to a redundant clock source of arbitrary phase produces frequency swings.

Between nodes the frequency accuracy should be 1×10^{-11} or better.

PAGING CONTROLLERS

Paging systems are generally composites of a controller from one manufacturer and a transmitter from another. There are exceptions, and a small number of manufacturers supply complete systems. In principle, almost any digital compatible RF transmitter with a duty cycle approaching 100 percent can be used for paging. The way it works and the features provided will depend almost entirely on the controller.

Controller manufacturers often specialize in a part of the market only. Some produce small capacity systems, some large and some (like Unipage) make extensive use of modularity to produce a unit that can be initially installed as a small system but can be expanded to meet demand.

The Unipage Model 15 EX is available as a 3000-line controller, with one channel of RF, two trunk channels, and a voice messaging system for $17,000. The unit, which stands about one-quarter rack high, is expandable to two RF channels and 16 trunks. However, the cards can be reused in the larger frame configuration to a total of 2 million subscribers, 1920 trunks, 128 RF channels, and 256 modem ports. Part of the secret of expandability is extensive use of distributed intelligence. This means that an expensive, high-powered central processor is not needed in the small systems.

REDUNDANCY

An important part of the controller is functional redundancy, which prevents loss of calls in the event that part of the system fails. Most redundancy operates on a 1+1 basis, that is each functional part has a counterpart hot standby that is ready to take over immediately if an error is detected. This necessarily requires a high level of internal checking, so that data or processing errors can be identified quickly and counteracted. However, with larger systems redundancy can safely be changed to N+1, meaning that every part has a redundant counterpart but that there may be N active units. This is much cheaper than full redundancy (or 100 percent 1+1).

Redundancy on a budget is possible by duplicating only critical units such as processors and allowing the occasional line card to fail on the grounds that only a few calls will be involved. In fact, this is commonly done in small systems (less than 5000 lines).

Redundancy is also essential for routine and corrective maintenance, as the users of paging services include emergency personnel and others who regard reliable service as vital. When errors are detected in the component cards, it is usual that the faulty unit will automatically be taken out of service and an alarm will come up. Maintenance consists of replacing the faulty card, a process that can be undertaken without affecting service. It is an important plus if all cards can be safely changed without the need to power off first.

Networking

Large systems will require networking of controllers to allow wide-area coverage. It is usual that wide-area operators will offer service over a number of zones, including an "all zones" option. This mode requires cooperative interaction of all controllers, which may simulcast on all transmitters or may use a queuing procedure to pass the call around the network.

Networking on a nationwide basis may involve access to a satellite link, interworking and interfacing with other operators (particularly true in the U.S. where regional operation is most common). The satellite calls will usually be batched (to ensure a reasonable efficiency) and sent via a dedicated link or switched PSTN line to an appropriate earth station.

Alpha-Numeric Direct Access

Although most alpha-numeric messages are passed through an operator, direct access via a modem on a DID (direct indial) trunk can be provided as shown in Figure 10.21. An industry standard protocol, TAP (Telocator Alpha Protocol), is available for this link.

FEATURES

A number of features can be provided by the controller that can enhance the service offered to the user. These include the following.

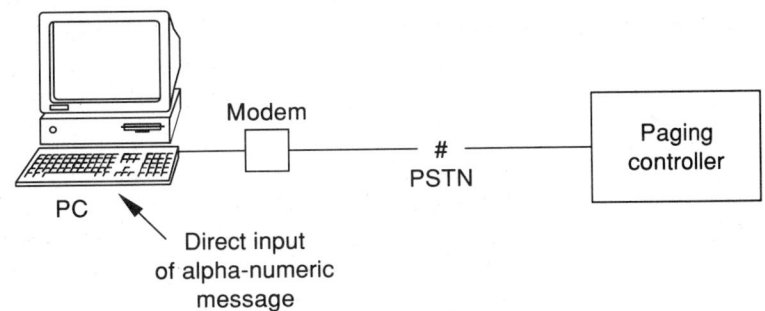

Figure 10.21   *The direct path for a user to access the alpha-numeric messaging via a modem.*

Tone Sub-Address Access

The paging tone only POCSAG format supports four different tones, which can be assigned by the user to indicate different things. For example tone 1 may signify the office, tone 2 home, tone 3 urgent, etc. One way to access these tones is to assign a unique telephone code to each tone. This is effective but wasteful of code. DTMF overdial can be used so that, for example, the user may overdial the digits 1, 2, 3, or 4 after connection has been made to the controller to indicate which sub-address is required.

Numeric Correction

Numeric messages are usually directly input via DTMF overdial by the user. Should the user incorrectly key the message (something that is easy to do particularly as there is no display feedback from most phones) a correction code can be provided. Errors detected in this way can be corrected without the need to hang up and dial again.

Alpha FAX

Alpha messages can be diverted directly to a fax machine. Rather than send numerous short messages, they are best stored and sent as required.

Group Call Merging

Early in the history of paging, the group call facility was introduced. This enables pagers to be called individually or as a group, depending on the caller. Recently it has become possible to group call any group of pagers, which may not even be of the same type.

Call Limit

Sometimes pagers will be given out as demos or loaned to clubs or organizations for short periods. To control the use of these devices, call limits can be placed on the pager, which automatically cancels its own validity when that number of calls have been completed.

CHAPTER
11

TOWERS AND MASTS

Towers (self-supporting structures) and masts (guyed structures) cost about the same if they are both short (that is, at heights ordinarily encountered in paging). As the structures get higher, the costs of guyed masts tend to increase linearly (for the same cross-section); the cost of self-supporting towers increase exponentially. Because guyed masts require a good deal of land, they are used mainly in rural areas.

In paging, tall towers are not usually needed and, except for rural areas, poles should be adequate for heights up to 30 meters.

Figure 11.1 shows the amount of land needed for towers of different sizes. Table 11.1 shows the amount of land needed specifically for three-leg and four-leg towers. In this table, T and W are the land dimensions used in Figure 11.1.

No structures of any kind should be built closer to the tower than the edge of the boundaries defined in Table 11.1 because the support of the surrounding soil against turning moments may be diminished. These dimensions are a guide only; the design of a tower or mast depends on such factors as wind loading, local building codes, and local planning-authority regulations.

The choice of monopole, mast, or tower is often made for the operator by the local-government rules or environmental considera-

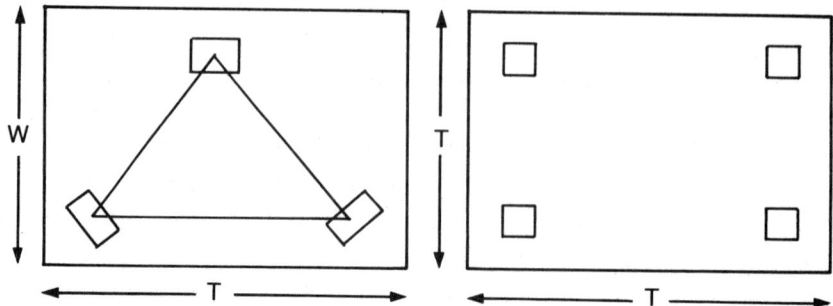

Figure 11.1   *Dimensions of land for towers of different sizes.*

Table 11.1   *Land usage and weight for three- and four-leg towers*

3 LEGS				4 LEGS		
TOWER HEIGHT	T	W	APPROX WEIGHT (tons)	TOWER HEIGHT	T	APPROX WEIGHT (tons)
10 M	7	7	0.7	20 M	7	1
20 M	8	7	1.7	30 M	9	2.2
30 M	10.2	9	3	40 M	10	4
40 M	11.5	10	6	50 M	12	8
50 M	13.8	12	10	60 M	13	12
60 M	15.5	14	14	70 M	14.4	16

tions. Sometimes, however, there is a choice, so it is worthwhile exploring the alternatives.

The location of the tower needs some careful consideration. In most countries, it will be sufficient that the tower win local zoning approval (from the local authorities) and does not constitute a hazard to air navigation. Usually, unless the tower is particularly large or the proposed location is zoned residential, there will not be too many objections from the local authorities. However, they may require the planting of trees around the structure (and in some cases even a painting scheme that is more environmentally sympathetic, such as sky-blue or green.

Conflicting with the requirements of being inconspicuous will be the requirements of the aviation authorities that the structure be visible to aircraft on VFR (Visual Flight Rules) and that it not be an obstruction to any existing or future flight paths.

Generally if the structure is smaller than 50 meters (in the U.S. special conditions apply above 200 feet) and more than 10 km from any airport it is unlikely to be a problem. In any case it is a good idea to get a ruling from the local aviation authority on both the location and warning markings/beacons that are required.

Information required by the aviation authorities will include

- accurate coordinates of the tower location
- structure height
- structure type (tower/mast/pole)
- proposed warning beacons and hazard warning paintwork
- location of the nearest airport and other airports within 10 km

Also it may be required in some countries (as it is in the U.S.) that a full inventory of the RF facilities—including the frequency, power, and radiation patterns—be provided. These are sometimes needed to assist in the evaluation of potential interference to IFR (Instrument Flight Rules) navigation equipment.

In the U.S. it is *compulsory* to get an FAA "determination of no hazard to navigation." This can take about 60 days, or longer if the submission is not complete.

MONOPOLES

In general, a monopole is more aesthetically pleasing, although very few neighbors are likely to welcome any support structure. The monopole, like a building, has a fixed platform height and usually comes in a very limited range of sizes (typically, 15–50 meters). It may have an internal ladder and cable tray. Its main structural advantage is the small land area required, typically 9–16 square meters (3–4 meters square).

Monopoles can be erected in about one day provided the base has to be poured and cured. Because they are available in a limited range of sizes, they often can be ordered almost off the shelf and have much shorter delivery times than towers.

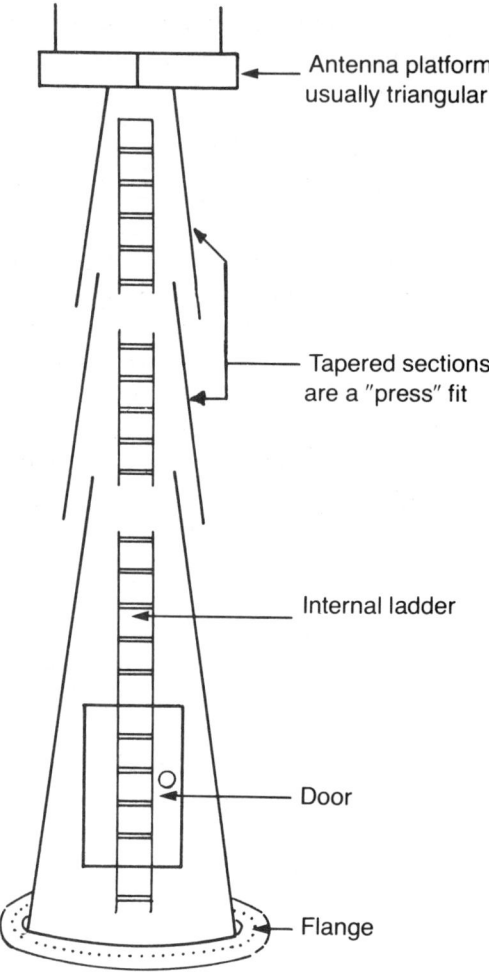

Figure 11.2   *The construction of a monopole with an internal ladder as used by Telecom, Australia.*

Monopoles are usually fabricated in tapered sections of about 10 meters apiece and fit together simply by stacking the sections (see Figure 11.2).

The support is simply a cage, designed to withstand large turning-moments, embedded in concrete. Shafts typically 8 meters deep and tapering from 2 meters at the bottom to 3 meters at the top form the foundation. Other monopoles can have much wider bases with correspondingly smaller shafts. Bolts 2 meters long and 57 millimeters in diameter, embedded in the concrete, attach to a flange at the bottom of the shaft. Up to 50 bolts can be used to hold the flange.

These structures can be designed to give the torsional stability required for low-frequency microwave bearers (maximum 1/2-degree twist).

GUYED MASTS

Guyed masts are practical only where land is inexpensive. They often prove to be the cheapest solution in rural environments.

Guyed masts are usually constructed of sections of triangular cross-section about 6 meters long. The sections are typically 0.5 to 1 meter wide per side and are designed to be bolted together. The strength of a mast is essentially in the guying, so proper tensioning of the support cables is vital. Concrete anchors hold the guy wires. For paging applications, the standard cross-sections should prove adequate to accommodate the paging and microwave link antennas. Figure 11.3 shows a 50-meter mast in Asia.

Figure 11.3   *A 50-meter mast in Asia. Notice the long grass (a fire hazard) in the foreground and recent evidence of fire in the very near foreground.*

Figure 11.4   *A hybrid mast/tower constructed on a rooftop.*

Hybrid Structures

Sometimes it is not easy to decide if a structure is a mast or a tower. Figure 11.4 shows a structure that started as a tower and later sprouted a mast on top. This structure is on a rooftop in Manila, the Philippines. The mast may have been built to reduce the total loading on the roof.

Fabrication

The cost of a mast is related more closely to the weight of steel than to the height. Table 11.2 shows that the weight of a mast is almost linearly dependent on its height. It can be seen by comparing Table 11.2 with Table 11.1 that self-supporting masts increase more rapidly in weight than guyed masts, particularly for structures higher than 30 meters.

Table 11.2 *Mast height and weight*

MAST HEIGHT (meters)	APPROX. WEIGHT (tons)
30	1.5 to 3
50	3 to 5
70	4 to 7
100	10 to 12

TOWERS

Towers are self-supporting structures that are most practical when land is expensive. Figure 11.5 shows a tower that supports a number of microwave dishes (the solid dishes) and gridpacks (the wire-formed dishes).

Figure 11.5 *Some towers are designed for high-density dish loading. Notice that each of them has a cover for weather protection.*

Towers require less land than guyed masts and are capable of supporting a large number of antennas, a factor where you plan to rent tower space to other users. When other microwave facilities are planned (for example, for a wireline carrier), a self-supporting tower is probably the best choice.

A three-sided tower is usually the best value (load-carrying ability per dollar). For the same strength, however, it has a wider base and requires more land than a four-leg tower. A four-sided tower has an extra face and can carry more antennas. For a given strength, it is also smaller.

Towers members can be of various types, including solid, tubular, and channel sections. Tubing is the cheapest material for tower construction; it is available in a large number of sizes and needs little work to make it suitable for towers. Tubing does, however, have a long-term maintenance liability. Moisture can build up inside the tubing and cause corrosion, and in extreme environments, can freeze, thereby splitting the tubing. In coastal areas or in areas near heavy industry, this type of construction could prove to be a liability. Such towers need to be designed with weep holes, and the holes need periodic cleaning and unblocking.

Round-bar members can be made of round solid sections; they do not have the corrosion problem of tubular sections. They do, however, require substantially more steel for the same strength and thus weigh and cost more.

The most common material used in tower structures is the channel section, which can be made from formed-plate or angle sections. Formed plate is cut from rolled plate; it is cut to length with the bolt holes punched while still in the plate form. It is then cold-formed into 60-degree or 90-degree channels along the center axis.

Deformed plate is made from milled 90-degree sections. For 60-degree sections, the plate is bent another 15 degrees on each flange. This plate is cheaper than formed plate, but often it is not precision-formed, which can lead to problems with bowing.

SOIL TESTS

Before a tower, mast, or pole can be erected, it is necessary to conduct a soil test. This involves taking core samples of the ground on which the structure is to be built and then having the samples analyzed. Using the results, the design engineer can determine the load-bearing capacity of the soil and its ability to resist the turning moments of the footings. Only after this test is complete (from one to four weeks) can design of the structure foundations begin.

OTHER USERS

If the structure is in a particularly prominent position, you should consider, before the tower is designed, the prospect of obtaining additional revenue from leasing tower space to other users. A modest increase in cost at the design stage can significantly improve the load-carrying ability of the structure.

When planning for other users, include them in the overall design by assigning their number, antenna type, and positions on the tower at the design stage. The structural design should also include detailed drawings of the proposed positions of other users so that they can be allocated at a future date without the need for new load calculations.

In general, operators need not fear that including other users will cause interference, however the converse may not be true.

Other users' services sometimes sprout up almost spontaneously in certain prominent areas and are known in the trade as "antenna farms." These "farms" can appear almost anywhere; Figure 11.6 shows one such farm thriving on a rock face in Baguio, in the Philippines.

Figure 11.6   *An "antenna farm" on a rocky outcrop. These antennas are mainly TV antennas mixed with a few links. Notice that the high-gain antennas are often mounted off-vertical, where they won't work well.*

TOWER DESIGN

The antenna structure must be designed by a structural engineer, but it is worthwhile to consider the design parameters. The structure must account for gravity loads (dead loads) that include structure weight, antennas, and ice, as well as live loads, such as those caused by wind and seismic activity. Invariably, wind, ice, and tower fittings will provide the dominant loads on the tower.

Wind Loads

Until very recently, the dynamic load caused by wind was not fully understood, and towers were designed to withstand a known static load, which was increased by a safety factor (often doubled) to account for dynamic effects. In the light of recent studies, it is clear that early designs tended to overdesign the bases and underdesign the top portions of the structures. Particularly in typhoon and hurricane areas, the top portions of old designs are now being strengthened.

In general, as more collapsed structures are studied and more detailed information of long-term wind peaks becomes available, the minimum requirements of codes for determining wind loads have consistently increased. Old structures should therefore be used only after a thorough survey and inspection.

Wind speeds, recorded by national authorities, are of interest to a tower designer. The designer should know the peak gusts (instantaneous readings) and fastest-minute-wind (the highest velocity sustained for one minute). These two figures are connected by a ratio of approximately 1.3:1.

Fifty-year peak wind velocities are sometimes interpreted as ones that are expected to occur 50 years apart. This is not an accurate interpretation. A better interpretation is that 50-year peaks are ones that occur with a probability of 2 percent each year. Therefore, the fact that an old tower is still standing may simply be good luck!

Typical Specifications for a 40-meter Tower

The tower designer must know a number of things about a tower before beginning the design process. The following lists contain the considerations required for a typical 40-meter tower:

- Four-sided (or three-sided).
- 40 meters.

- Designed to EIA RS222D (the U. S. standard), Australian Design Standards, or other preferred standard.
- Stress factor (that is, suburban or rural safety factor). For example, in the Australian design code for suburban areas this stress factor is:

  $1.7 \times$ factor on steel.

  $1.75 \times$ factor on foundations (this factor can be found from the relevant design code).

- Zone specifications and wind loading, depending on location.
- Maximum allowable twist (0.25 degree for 7-GHz microwave or 0.15 degree for 10 GHz) (mainly for links).
- Maximum allowable tilt (1 percent for 7-GHz microwave or 0.5 percent for 10 GHz).
- Platforms and walkways at the levels where access to microwave dishes will be required.
- A platform of about 2×2 meters at the top, with guard rails 1.5 meter high and suitable for mounting paging antennas at the edges. The mounts will be used to attach antennas with tubular supports up to 70 mm in diameter using three heavy-duty clamps.
- Cable tray will be accessible from the ladder and will be 0.6 meter wide.
- Safety guard around the ladder, which will be internal with respect to the tower. (See Figure 11.7.)
- IAO standard paintwork and an aircraft warning beacon at the top.
- Tower orientation.
- Specify the microwave dishes, type (solid or grid pack) and mounting level. Allow for future expansion (even if expansion is not planned, it will probably be required; a good rule is to estimate the future requirement and then double it).
- Tower footings should be confined within a square plot of land (as specified earlier in this chapter).
- Provision of lightning arrestors.
- Grounding of tower on all four legs.

Figure 11.7   *When towers are to be climbed by staff other than riggers, safety guards should be provided.*

SECURITY

Towers are attractive to youngsters, who see them as a challenge to climb. If towers are climbed untold damage may be caused to the antennas and cables, and even worse the youngsters may suffer serious injury or death as the result of a fall. In order to discourage unauthorized access, a human-proof fence should be installed around the tower (as shown in Figure 11.8) or access can be barred by the attachment of spikes around the legs (as seen in Figure 11.9). In all cases it is advisable to place a notice, similar to the one shown in Figure 11.10, on the base of the tower to deter trespassing.

HOW STRUCTURES FAIL

A free-standing structure such as a tower is most vulnerable in the compression leg (the side away from the direction of the wind). A mast is similarly subject to compression failure, but because of the

Figure 11.8   *A human-proof fence (with spiked steel posts) around a rural cellular site.*

Figure 11.9   *Access can be restricted by the use of spikes on each leg and on the access ladder.*

Figure 11.10   *A warning sign at the base of a tower in England (this is on the same tower as shown in Figure 11.9).*

multiple guying points, has a more complex failure mode. Failure in both instances will probably be due to buckling.

The stress is very sensitive to wind velocity. It varies as the square of the velocity for static loads and as the velocity to the power of approximately 2.5 for dynamic loads. Wind speed varies more or less regularly with height and has an approximately parabolic gradient from ground level to 400 meters.

A less predictable factor is turbulence, although this is probably the major factor in structural failure. Turbulence is poorly correlated along the length of the structure (it is randomly distributed) and varies rapidly with time. In modern studies, the very unpredictable nature of turbulence is taken into account, and it has been found that some turbulence patterns are significantly worse than others.

Topology plays a part and many large towers will be situated on hilltops to gain additional elevation. Hilltops unfortunately produce increased airspeeds over their crests and a 10-percent hill slope can produce a 20-percent increase in airspeed or a 40-percent increase in wind loading. This is the reason that windmills and wind generators are usually placed on hilltops.

Stiffness (the ability to resist deflection) is a sought-after characteristic in structures and an important factor for reliable microwave operation. Stiffness is often obtained, however, only by using more metal, which increases the cost and weight. For economic reasons, modern structures are designed to minimize the amount of materials used, so a trade-off occurs. Adding extra dead loads (for example, equipment and antennas) reduces stiffness.

TOWER, MAST, AND MONOPOLE MAINTENANCE

The unscheduled replacement of the antenna support structure can be both costly and disruptive to the installed service and should be avoided if at all possible. The collapse of a tower or mast, particularly in a populated area, can be at best embarrassing and at worst a catastrophe.

Antenna support structures require regular routine maintenance, which is often neglected on the grounds that the structure has been up for years and has not shown signs of fatigue to date.

To appreciate the need for competent inspections, it is necessary to first understand how and why structures fail. These are the major causes of failure:

- Poor design, which inadequately allowed for static or, more frequently, dynamic wind loads
- Overloading of the structures with too many antennas and feeders
- Corrosion, particularly where hollow structural members are used
- Insufficient attention to guy-wire tensions and conditions (corrosion)
- Inattention to the indicators of stress
- Guys corroded or improperly tensioned

As an interesting example of the problems facing masts, the mast shown earlier in Figure 11.3 and again in Figure 11.11 is worthy of a closer look. The long grass in the foreground of Figure 11.3 represents a fire hazard, and evidence in the extreme foreground indicates a recent fire.

This mast uses passive reflectors (the large plates at the top) to deflect a microwave link to the ground-mounted receiving dishes

Figure 11.11   *The dishes with conical radomes (to help water run off) mounted at the base of the tower shown in Figure 11.3. Notice the spalling concrete base.*

illustrated in Figure 11.11. These dishes are protected by conical radomes. The cracking and spalling concrete seen in Figure 11.11 at the base of the mast is a sign of excessive stress.

Masts are held up by guy wires that are anchored into concrete blocks. Signs of stress were evident at all of the anchor points at the site in Figure 11.11. Figure 11.12 illustrates cracking and spalling at these points at this site. All of the anchor points inspected on this structure showed signs of spalling. This mast was well painted and relatively rust-free, but as Figure 11.13 illustrates, little attention was given to mechanical details. The buckle linking, the guy wire to the anchor point in Figure 11.13, shows that the bolts were not tightened and washers were not used. The large central bolt is about 40 mm in diameter.

Routine inspections should be carried out about once a year for structures located near the coast and every two to three years at sites more than 100 km from the sea, as well as after severe storms or periods of prolonged heavy icing.

Figure 11.12   *Cracking and spalling of the guy-wire anchor blocks.*

Figure 11.13   *The buckle connecting the guy wire to the anchor block in Figure 11.12. Notice that the bolts were not tightened and that no washers were fitted. The structure was, however, relatively rust-free.*

INSPECTION

Very few paging companies are large enough to employ a full-time, qualified structures inspector. Those that can will invariably be wireline operators.

Because of the special nature of support-structure maintenance, the paging operator will generally find that there are few companies with the necessary expertise and that the availability of those companies is limited. Having found a competent operator, it is therefore a good idea to arrange the maintenance on a contract basis. The company should have a good structural engineer and experienced inspectors who can climb and inspect every portion of the structure. The inspection process should begin with a review of the existing documentation about the structure and its fixtures. It should then proceed step be step, using a checklist like the one provided at the end of this chapter.

If only paging, cellular or mobile two-way (PMR) antennas and microwave links are mounted on the tower, the inspection can be carried out without disturbing the operation. The inspector should avoid prolonged periods of exposure (more than 10 minutes) within 1 meter of the antenna. The relevant local RF radiation limits should be observed.

STIFFNESS

A structure that is too flexible is subject to excessive stress and is liable to failure. All structures have resonant modes about which they vibrate. The primary mode for a free-standing tower involves its whole length and results in maximum movement at the top. The tower will sway under wind loads and the period of this sway is a measure of its stiffness. This period is the time to complete one full cycle (that is, from the vertical position through to the maximum deflection and back to the vertical is one half a period). This period can be measured by observation (difficult and inaccurate), by a video camera (better), and by an accelerometer (best).

Accelerometers are usually located at three or more positions along the length of the structure; the results are relayed to the ground for later analysis. Equipment records motion in two directions, as well as torsion. The optimum period is a function of the structure height, strength, design, and mass. For a 180-meter tower, a two-second period is good; a four-second period would indicate excessive flexibility.

Because early design codes did not fully appreciate the effects of dynamic wind loading, underdesign of the top portions of the structures was common (together with overdesign of the lower portion). As a result, the flexibility of the top portions often, over time, causes high levels of stress. Strengthening the top portions is thus often required at a cost of approximately 10 percent of the structure cost. Operators will probably encounter this problem only if they use an old, existing tower; design techniques today properly account for the distribution of stress. A good indicator of stress is localized flaking paint and, in some instances, corrosion. Flaking paint is best detected soon after a storm when the recent stress highlights the problem.

REPAIR

Any repair and maintenance indicated by inspection should be undertaken as soon as practical. Finding suitable contractors to do the work may be difficult.

Towers should be painted once every five to seven years, depending on the environment. Painting and touch up for corrosion can be done by many contractors, particularly by those who specialize in heavy industry or bridges.

Replacing bolts, adjusting antennas, and low-stress members can be done by a suitably qualified rigger.

Stress problems are more serious, however, and require the intervention of a structural engineer. The stress problems could be due to weakened members but are more likely design-related. After analysis, the structural engineer can recommend the necessary modifications. The replacement of high-stress members requires the services of a specialist structures contractor.

Stress can be reduced by lowering the wind loading of attachments, but it more often involves adding structural members. The structural engineer usually considers various alternatives to reduce stress and recommends the most cost-effective one.

Welding of strengthening members often destroys galvanizing and other protective coatings, so protective coats will be needed.

TOWER INSPECTION CHECKLIST

A tower inspection should include the following steps:

Tower

1. Check foundations, ground points, and straps.
2. Check for corrosion and condition of painting.
3. Check welds for cracks, using ultrasonic equipment where necessary.
4. Check for signs of stress, particularly flaking of paint or bowing of members.
5. Check all bolts for proper tension and corrosion. (Some may actually be missing).
6. Check guys for proper tension and possible corrosion. In some areas, anticorrosive agents must be applied.
7. Note the position of all fixtures, and when these positions differ from the records, note the details (including photographs).
8. Check for bent or fractured members.
9. Check for tower twist or distortion (sometimes twist can be detected by checking that the tower lines are true).
10. Check the condition of the galvanizing.
11. Check for corrosion in hollow members; this can sometimes be detected by hitting members with a hammer and listening for falling rust. (In some instances and particularly in corrosive environments such as along the coast or in heavy-industrial areas, a low-stress member can be removed and replaced by a new one. This member can then be examined in a laboratory for strength and corrosion.)
12. Keep a permanent log of the inspection.

Grounding

13. Check that all clamps and ground straps are secure and in good condition.
14. Check that bolts are covered by an anticorrosive material.
15. Check that the lightning rod is secure and in an effective position relative to all antennas (higher than any antenna and at least three wavelengths away). (Some installations dispense with lightning rods and use DC ground antennas instead. This

is sometimes essential where space on the tower top will not allow for a reasonable separation between the antennas and the rod.)

Antennas

16. Check that all antennas are vertical or at the correct angle of downtilt.
17. Check the physical condition of the antenna; it should be free of cracks, dents, and burns.
18. Check that all bolts and clamps are secure.
19. Check that the antenna grounding is secure.
20. Check that the feeder grounding is secure.
21. Check that the feeder support is adequate and not causing wear or fatigue.
22. Check for any slippage of the feeder.
23. Listen for any audible signs of gas leakage in pressurized systems.
24. Check that the antenna "tail" connector is properly sealed.

Anchorage and Foundations

25. Check that concrete anchors are free of spalling (flaking) or cracks.
26. Check that anchor bolts are tight.
27. Check that grounding is secure.
28. Check that anchor rods are not rusted or corroded.
29. Check for any signs of anchor slippage or creep.

Guy Wires

30. Check for any signs of rust or broken strands.
31. Check that the connectors to guy wires are in good condition.
32. Check that the turnbuckles are in good condition.

Tower Lighting

33. Check that all beacons are in working order.
34. Check that all beacons are in good condition.

35. Check that beacon drain holes are clear.

36. Check that beacon reflectors are in good condition.

37. Check that beacons are free of signs of moisture.

38. Check that beacon lenses are clean.

39. Check that beacon wiring is in good condition.

Ice Shield

40. Check that the ice shield is secure and undamaged.

 # POWER
AND PROTECTION

There are many times when it is necessary to consider power conditioning. The billing system must have a reliable and continuous power supply as must other computers that are handling databases such as subscriber lists and customer data. A power failure that occurs while data is being manipulated can cause corruption of the database and perhaps even the irretrievable loss of valuable information.

Computers are notorious for their unreliability in areas where the consistency of the power service is poor. This is largely because most computers use switch-mode power supplies, which offer the circuitry that they power very little protection against power surges.

The most common form of backup for a computer is the UPS (uninterruptable power supply). This is usually a device consisting of a rectifier that charges a small sealed battery, which in turn drives an invertor. These devices usually have ratings of around 900 watts to 2000 watts, and are rated for approximately 30 minutes. Larger cabinet sized units may have capacities of up to 15 KVA for four hours. As is illustrated in Figure 12.1, the conventional UPS is a simple double-conversion device where the incoming AC is converted to DC, which in turn is converted back to AC.

Such a device is free-running so that in the event of a power failure the battery continues to provide the power to run the computer or switch. For typical computer usage, a backup time sufficient to finish

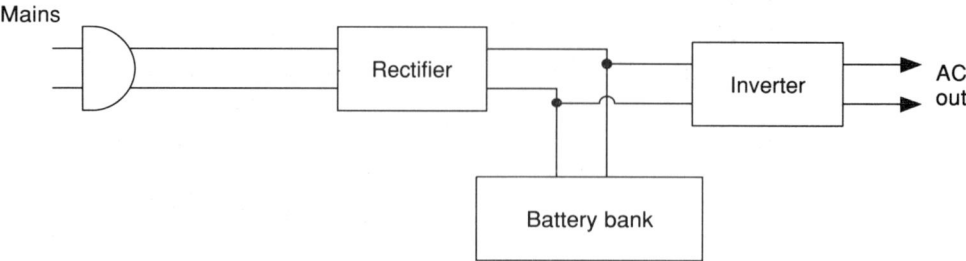

Figure 12.1   *A typical UPS configuration.*

the current job in such a way that data will not be lost as a result of the power failure is all that is needed. Other equipment and the billing computer may be required to operate even during sustained power breaks and may therefore require more substantial backup times.

Since the efficiency of the typical double-conversion UPS is not high, (being typically 60 percent at full load), the efficiency is very load-dependent, decreasing rapidly at lesser loads. You should not rely on the implied ampere-hour capacity and so assume that for lighter loads the UPS will run longer. In practice, for most UPSs when run at 50 percent load you cannot expect much more than a 10 percent increase in backup time.

Big double-conversion units (more than 5 KVA) also tend to be noisy, to the extent that they cannot be placed in the vicinity of the workplace. Also, the low efficiency means high heat losses, which can be a strain on the air conditioning. The heat load can be calculated directly from the efficiency, because the losses are all dissipated as heat.

FERRO-RESONANT UPS

A more efficient, but more costly solution is the ferro-resonant UPS, which uses a saturated resonant transformer to clean up the incoming power line. It uses a fly-wheel effect to overcome short duration voltage sags, and inherently dampens spikes, surges, and noise.

The ferro-resonant transformer can supply clean power for around 15 ms after loss of the primary power. In a properly designed system this will allow time for the standby inverter to start up and take over the supply. This is all accomplished without any noticeable loss of power, but it means that the inverter is only running when it is needed. Because of this, ferro-resonant UPS is much more efficient than the double-conversion system. Efficiencies of 90 percent can be expected and so the heat load is much reduced. Noise is a lesser problem and the inherent filter characteristic of the transformer eliminates the need for additional filters. The inverter, which is only run on a needs basis, will have a longer life and increased reliability.

Ferro-resonant conditioners are based on a constant voltage transformer principle. They are simple, reliable, and perform well in practice. They are (as the name implies) resonant and so are dependent on the line frequency. Exporters should note that both 50 Hz and 60 Hz main power supplies are commonly used and that a ferro-resonant conditioner will not work (without modification) at the wrong frequency.

UNINTERRUPTABLE BATTERY SUPPLIES (UBS)

This is a relatively new concept based on a combination of a ferro-resonant UPS, with the batteries backed up by a DC generator.

POWER STANDBY UNITS

These are similar to the UPS but with one very important difference, which is that they operate on standby and only come into service when an actual power failure occurs. The time to come on-line is usually hundreds of milliseconds, and so this type of equipment is not suitable for protecting operational computers.

Power standby units come in sizes up to 15 KVA. They are often powered by standard vented (wet) lead acid cells and so require monthly maintenance. The maintenance must include inspection of the acid levels and the terminals, which need to be checked and greased.

The power output from these units often deviates significantly from an ideal sinewave and so if the intention is to power computers it will often be necessary to insert a line conditioner between the computer and the standby unit. This is necessary to prevent damage to the computer's power supply particularly during the change-over operation.

Sometimes power standby units can be reconfigured for continuous operation and so can operate as a UPS.

POWER CONDITIONERS

These devices are used to clean up the waveform and to limit the power excursions that can occur. There are basically two types. The electronic conditioners sense the line voltage and respond to fluctuations. As a result they do not have instantaneous response times, and they usually respond with step-function corrections.

DC–DC CONVERTER

Often it will be necessary to provide DC–DC conversion to allow 24-volt batteries to power 48-volt equipment and *vice versa*. Suitable converters are readily available in both 19-inch and 600-mm rack mounting. Depending on the equipment being powered and the consequences of its failure 100 percent redundancy, with suitable isolation between the converters may be needed. In general, converters are considered to be reasonably reliable but not sufficiently so that they can be used to provide power to service affecting modules like links or MUX on a stand-alone basis.

RECTIFIERS AND BATTERIES

Rectifiers are available in the conventional transformer-coupled style and as switch-mode devices.

The conventional rectifier, while at least two to three times the size and weight of the switch-mode device is much more reliable, particularly in areas where power conditioning is poor. This type of rectifier can be easily serviced as it contains no really complex components and its components can readily be sourced. The transformer itself provides good isolation from power surges and transients. The MTBF of a well-engineered rectifier of this type is around 10 years.

Increasingly the switched mode rectifier is becoming popular for telecommunications equipment, and it has the advantage of being so small and light that a rectifier bay, complete with sealed batteries can be positioned in the equipment suites in standard 600-mm bays. Not only does this enhance the appearance of the equipment and save space, but with attention to the placement of the rectifier bays, it is possible to make the DC power distribution losses very small indeed. Particularly when compared to the old-style battery room with its bus-bar distribution network, the savings in copper alone can amount to 20 percent of the total DC power costs.

On the down-side the complexity of switch-mode equipment is greatly increased by the fact that the supplies are virtually DC-coupled. This means that in the event of failure of part of the system, repair is nearly impossible without access to dedicated repair facilities. Because of this, manufacturers of this type of equipment usually limit the internal construction to two or three boards and recommend that local repair be on a board replacement basis. At least one manufacturer has recently produced a single-board rectifier so that repair simply amounts to replacing the board. When purchasing switch-mode rectifiers it is essential to order adequate supplies of spares and to be comfortable with the supplier's arrangements for component-level repair.

Whichever type of rectifier is used, it should have load-sharing capability so that the load is automatically shared between the available rectifiers.

The rectifiers have alarms for such conditions as power failure, low output, and rectifier failure. These outputs can be wired to the spare alarm positions in the base station so that they can be remotely monitored.

AC power can be provided either as single phase or three phase. If a three-phase supply is used, the load can be distributed (as equally as possible over the three phases). This must be done carefully because most base-station hardware is single phase.

Rectifiers can be supplied in rack sizes compatible with the cellular equipment. Most equipment on the market, however, is not rack compatible. They usually come in modules of 25, 50, 100, or 200 watts (single phase), and somewhat larger in three phase.

Because rectifiers are relatively bulky (in terms of weight and floor space), plan ahead for the floor space needed. It should be possible to get about 200 amps single phase in one 600 mm rack, although this depends on the rack height. Plan your space carefully—for example, avoid buying rectifiers of about 100 amps that use just over half a rack height ("wasting" the rest).

POWER RATING

Rectifiers will be available up to about 150 amps DC in single phase, but units from 150 amps to around 1000 amps will be three phase. The DC output can usually be increased to any desired level by using parallel rectifiers. If this is done, automatic load sharing is essential.

Figure 12.2   *Three 100 amp conventional rectifiers working in a load-sharing mode.*

Ordinarily, two or more rectifiers are provided on a load-sharing basis, with provision to load share in the event of a rectifier failure (see Figure 12.2).

BATTERIES

Batteries are needed to keep the equipment functional during power failures. Sealed batteries are popular because of their low maintenance cost and flexibility in mounting arrangement. You can save space with sealed batteries because the need to partition the battery room is eliminated. Although some installers are content to place wet cells in an equipment room, the practice is not universally accepted. The initial capital cost of sealed batteries is, however, somewhat higher.

Some operators insist on having two battery banks in parallel (with half the reserve capacity each) to ensure against base station failure in the event of the failure of one battery bank. A single cell or even a fuse can cause battery bank failure.

You can calculate the rectifier load and battery load from heat loads

Rectifier load = sum of heat loads
Battery load = rectifier load + air-conditioning load
    (if included)

where

the air conditioners also run from a battery
the approximate battery load = 2 × rectifier load (that is, most electrical energy is dissipated as heat so the air-conditioning load is approximately equal to the rectifier load).

Some manufacturers produce equipment to temperature specifications that are sufficiently high so that air-conditioning is not essential during power failures. This results in significant savings in both batteries and emergency plant costs. It, however, results in a substantial temperature cycle for the equipment, which has been shown convincingly to increase MTBF.

Battery ampere-hour ratings are usually quoted for a 10-hour discharge. A paging base is ordinarily equipped with 2–3 hours battery reserve; at these higher discharge rates, the battery ampere-hour capacity is reduced by 10 to 20 percent.

EMERGENCY PLANT

It is sometimes necessary to provide a diesel generator set to back up the base station. When such a generator plant is provided, it is possible to reduce the battery capacity to the point where it merely ensures continuous operation from the time of power failure until the generator starts. Ordinarily, the generator starts automatically when a power failure occurs.

Generators for paging base stations will usually be single phase but may be three phase, depending on the main power supply. The generator is usually a diesel and with proper maintenance can be expected to give 10,000 hours of use if it is low revving (below 1800 RPM) and about 5000 hours if it is a higher speed unit. In many applications, the operator may rightly consider that the generator will receive only very light use, and the prospect of installing a used unit may arise. Generally there are few disadvantages to doing this, and the savings can be considerable. A used unit with less than 1000 hours of service, which has been properly maintained, may cost half as much as a new unit and give virtually the same service life in the cellular environment. Naturally maintenance costs will be a little higher than could be expected for new machines, and there will be no warranty, and these factors must be offset against the capital savings.

Maintenance should include running the generator once a month, ensuring that the fuel is not more than 12 months old, and following the procedures recommended by the manufacturer. Full load tests, which may require the use of a dummy load, should be conducted annually.

Gravity feed tanks can be very dangerous in the event of a fuel-line failure, which could flood the equipment room, and are therefore not recommended. The preferred delivery method is via a pump. To ensure that sufficient fuel is available for starting the generator, however, place a small gravity feed tank in sequence with the main fuel tank. The main fuel tank is best placed underground with a pump for fuel delivery. Dual pumps with manual change-over are a good idea.

Because it is costly to move a generator plant, it is best to purchase a unit that can run a fully equipped base station from the start. However, diesel plants do not perform well at partial load, and, unless expansions are foreseen within a few years, it may be necessary to plan to upgrade the generator plant at a future time.

Generators are usually rated in KVA. Unless the power factor is known, a figure of 0.7 should be used. Thus the generator rating, in KVA, is:

$$\frac{\text{WATTS}}{1000} \times PF \times E_f$$

where

WATTS = the total power consumed by the base
        station
PF = power factor
E_f = rectifier efficiency (typically 7080 percent)

Such a generator consumes about 0.3 liters fuel/KVA/hr.

You should provide a fuel tank sufficiently large to provide a one-week backup (this can be tailored according to the reliability of the local power supply).

Diesel fuel does not keep indefinitely; do not store it longer than six months.

Because the base stations may consume hundreds of amps of current, the DC distribution system must be properly designed. In particular, it is important to provide switches that can isolate each piece of equipment used. This isolation of the batteries, rectifiers, and equipment bays is imperative. Heavy-duty copper cables, carrying around 10–100 amps each, are usually used for power distribution, with each RF rack being individually supplied via a separate fused path equipped with a circuit breaker.

Use cables for rack wiring that are sufficient to carry the current safely. Table 12.1 shows the current capacity of various wire gauges. As the ambient temperature increases, a derating factor must be applied to the cable. Table 12.2 gives the appropriate derating. Table 12.2 assumes that not more than three separate conductors are placed in one cable or raceway. When more than three cables are bundled together, a further reduction in capacity occurs, as shown in Table 12.3.

Table 12.1 *Copper cables and their dimensions and current carrying capacity at DC continuous rating at room temperature (30 degrees C)*

AWG	MAX. CURRENT IN AMPS	DIA (MM) 19 STRAND CABLE	DIA (MM) SINGLE WIRE
26	1		0.04
22	5		0.064
18	10	1.16	1.016
14	17	1.84	1.63
12	23	2.32	2.05
10	33	2.95	2.59
8	45	3.7	3.26
6	60	4.67	4.13
4	80	5.9	5.2
2	100	7.42	6.54
1	125	8.43	7.35
0	150	9.47	8.25
00	175	10.06	9.27
000	200	11.9	10.04
0000	225	13.4	11.68

Table 12.2   *Correction factors for higher temperatures*

TEMPERATURE (C)	DERATING FACTOR
40	0.82
45	0.71
50	0.58
55	0.41

Table 12.3   *Derating factor for multiple bunched cables*

CONDUCTORS IN ONE CABLE OR RACEWAY	DERATING FACTOR
4–6	0.8
7–24	0.7

CABLES

Cables used for carrying the DC supply should be adequate for the peak current expected and the maximum expected operational temperature. Table 12.1 shows the current-carrying capacity of various wire gauges at 30 degrees C.

As the temperature increases a derating factor should be applied as per Table 12.2.

A typical high power RF bay at a base station will require around 100 amps. If this is derated to 45 degrees C (to allow for air-conditioning failure) then a derating factor of .71 would apply. That is, the cable should be able to carry 100/.71 or 140 amps. This would make an AWG "0" suitable for wiring these bays. If a slightly higher temperature is anticipated an AWG "00" would be needed.

Since the main factor limiting a conductor's current-carrying capacity is heat dissipation, bunching conductors together will lead to mutual heating and so to effective derating. The derating factor to be applied to bunched cables is given in Table 12.3.

GROUNDING

In order to protect personnel and equipment is essential that all installations are properly earthed. A good ground will minimize damage from power surges, lightning strikes, as well as from noise and interference.

The golden rule of grounding is to avoid ground or earth loops. To do so it is essential that all earths are firmly bonded together by straps of adequate current-carrying capacity. A typical ground loop, and one that can be quite dangerous, occurs between an unprotected telephone and its remote connection to the PSTN.

As seen in Figure 12.3 a lightning strike to a distant switch will cause a local potential rise that can amount to several thousand volts. Provided all the hardware in the distant switch is properly grounded and those grounds are bonded, there should be no damage done at

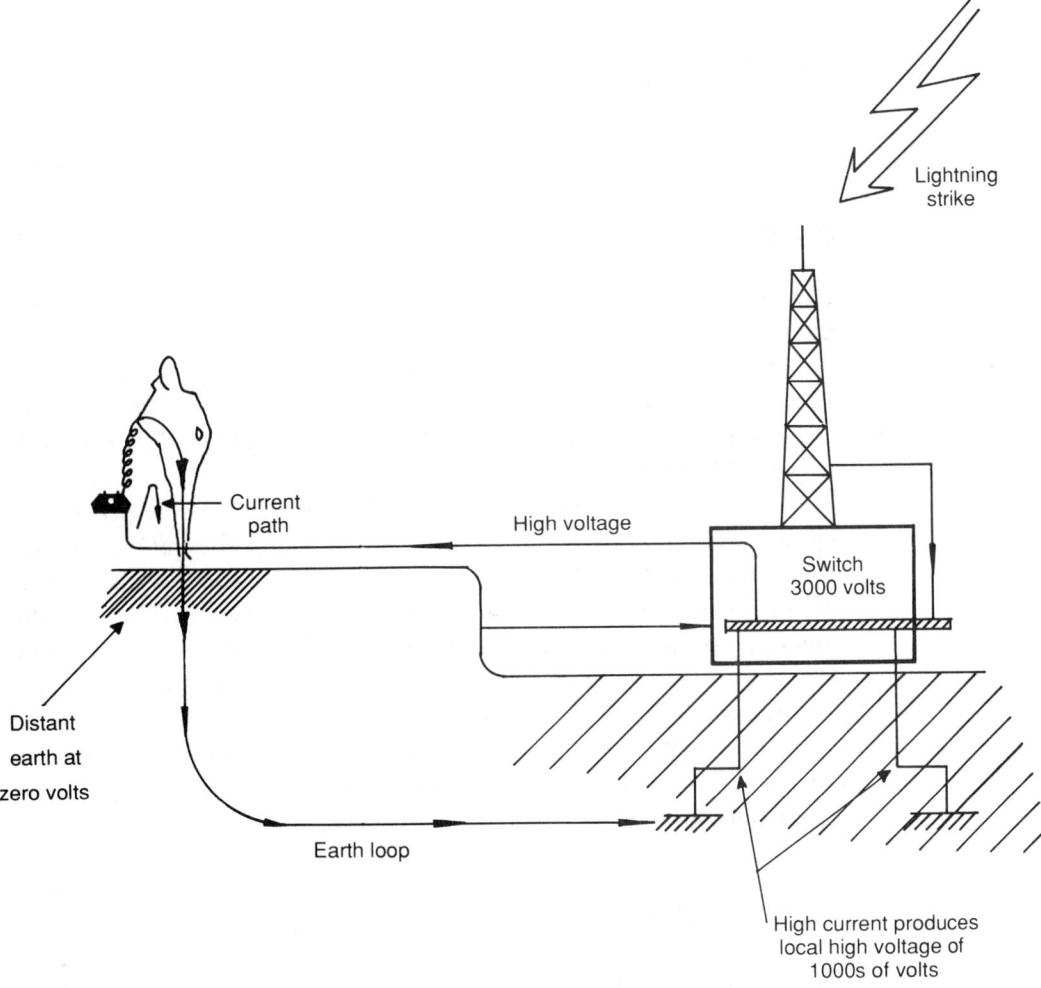

Figure 12.3 *An unintended and dangerous ground loop.*

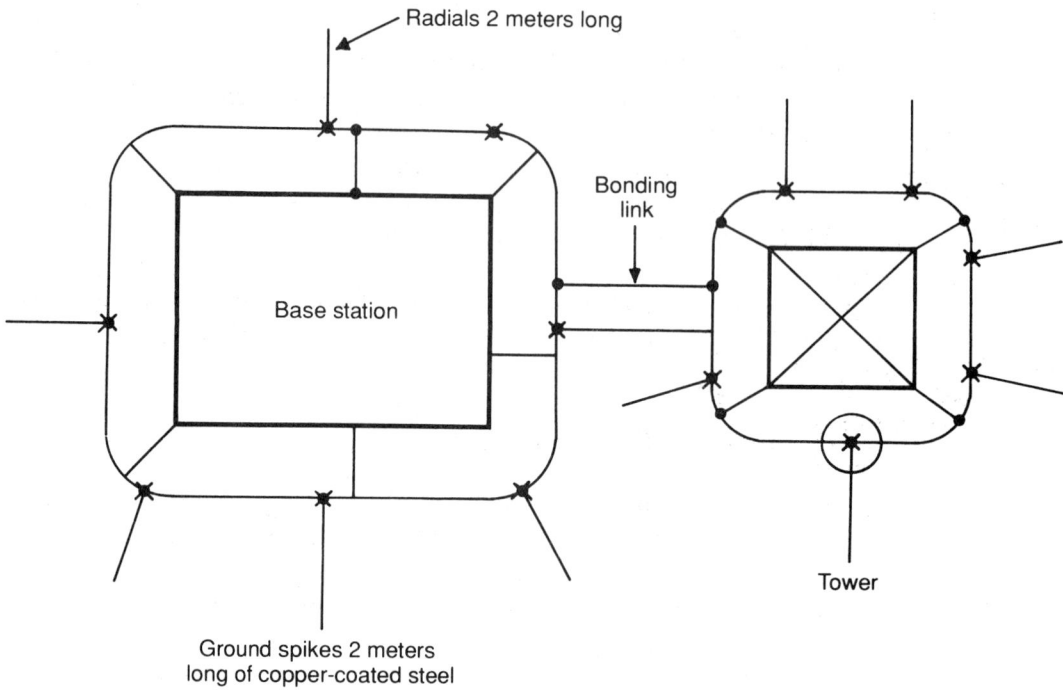

Figure 12.4   *A grounding ring. Note that the tower and base station rings are bonded together.*

the switch. However the user of an unprotected telephone line can provide an unintended ground loop with disastrous consequences.

Good practice at all switch and base sites is to install a ground ring around each structure, which is bonded to the ground by grounding rods, spaced at intervals of 2–4 meters (depending on soil resistivity). A typical ring grounding is depicted in Figure 12.4.

To ensure a low-resistance bonding, all joins should be exothermically welded, and tinned copper wire of gauge AWG 2 or larger should be used. Mechanically bonded joints should not be used for below-ground bonding.

Notice that it is very important to avoid direct connection of bare copper wire with galvanized steel as the combination will cause a serious corrosive, potential difference.

The 2-meter radials are there to assist the path to ground of large surges. The radials should be connected to 2-meter grounding rods at each end. The rods should consist of 2-meter, 15-mm diameter copper-clad steel rods. Stainless steel may be substituted if the rods are to be used near large mild-steel structures with which copper may cause corrosion problems.

The ring should be connected to the base station at 2-meter intervals by a gauge AWG 2 copper cable which is placed in a PVC pipe from the connection to the ring to a point at least 150 mm above the soil surface.

The tower should be connected to the ring at each leg in the manner described for the base stations. If a monopole is used, it should be connected to the ring at four different points around its circumference.

Avoid sharp bends since these can present a high inductance to surges. All bends should have a radius of at least 300 mm.

INTERNAL GROUNDING

An internal ground ring that completely loops around the inside walls of the base station at a height of 2 to 2.3 meters should be provided for internal grounding. Ideally this will consist of a flat copper bus-bar of cross-section 15mm × 5mm, otherwise a bare wire of gauge AWG 2 can be used.

This internal ring can be connected to a number of smaller bus-bars, which are designed as ground terminating points. They consist of a flat copper plate of approximate dimensions 250mm × 100mm × 8mm, and with holes drilled in them to make convenient connection points for straps to the equipment to be grounded.

A typical ground bar is depicted in Figure 12.5.

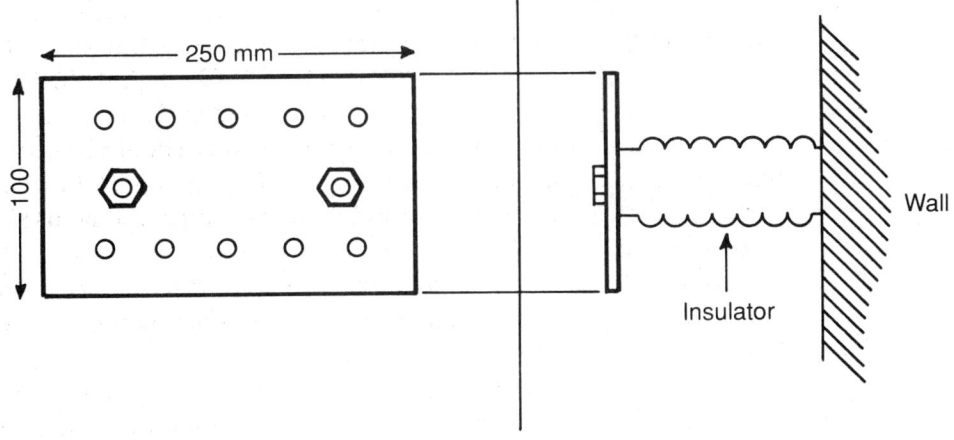

Copper ground bus-bar

Figure 12.5   *An internal ground bar.*

Figure 12.6   *An internal grounding bar cable*

All of the equipment racks should be connected to one of these ground bars using stranded AWG 6 copper wire. The racks should also be connected to each other by strapping each rack to the next. In order to avoid ground loops, it is advisable to preclude other grounds by insulating the racks from the floor. The cable tray should also be connected, preferably at a number of different points.

Grounding bars should be placed as needed and are usually drilled with spare holes for additional future connections as seen in Figure 12.6.

Cable Grounding

All RF cables should be well grounded at the top of the tower, the bottom of the tower, and at the point of entry to the building.

The cable window should have a grounding plate on either side so the cable sheath can be grounded at the external point of entry, and on the internal side there will be a good direct ground for the gas arresters.

All other cables such as telephone cables, interbuilding cables, and power cables need gas-discharge protection.

For telephone lines, the usual solid-state protectors should not be regarded as sufficient and three-leg gas-discharge devices should be used.

HIGH-RESISTIVITY AREAS

Where a good low-resistance ground proves difficult to achieve a "Ufer" ground can be considered. This consists of a cable embedded in concrete. It works on the principle that concrete is highly porous and will readily retain the moisture that it can absorb from the surrounding soil. An improvised Ufer ground can be obtained in low-rise buildings by attaching to the reinforcing bars of the concrete. To be effective a number of different connection points are usually necessary. This technique is of little use in buildings more than a few stories high.

Acceptable Ground Resistance

The resistance of the grounding should be less than 10 ohms (although 5 ohms would be preferable), when measured with a null-balance earth-resistance test set.

TRANSIENT AND SURGE PROTECTION

Irregular line voltage conditions can be caused by lightning, power system faults, electrostatic discharge, or radio frequency interference. Modern semiconductor components are extremely vulnerable to damage from two of these effects.

The most severe and most probable source of irregular line voltages in paging systems is lightning, which can easily find its way to the sensitive hardware by way of the antennas, power cables, or landline links.

The next most likely problem is electrostatic discharges. These can be the most insidious since the damage may not actually cause immediate device failure, but rather may lead to drastically reduced device lifetimes.

The magnitude of the problem for telecommunications companies has been increasing with the sophistication of the hardware used as can be seen from Table 12.4.

Protection Scheme

- Provide adequate lightning protection and good earthing.
- Protect all power lines connected to your equipment.
- Protect all RF cables at the point of entry.
- Protect all data cables and land-line entering the building.
- Eliminate ground loops.

There are a variety of protection devices available:

Power-Line Surge-Reduction Filters

These are placed in series with the lines and offer common and differential mode protection. They typically are rated at 100 kA per phase and are available in load ratings from around 10 amps to 1000 amps. The performance of is virtually independent of the actual load. It is essential that good grounding practices be followed.

In principle these filters are similar to the familiar π RF filter; a circuit diagram for one is seen in Figure 12.7.

In practice the problem is to have capacitors and inductors that can stand the high voltage and current surges. The inductor in particular has a tendency to fly apart under the stress of very high currents. Figure 12.8 shows a three-phase surge filter suitable for the protection of a base-station power feed.

Table 12.4

COMPONENT	DAMAGE ENERGY IN JOULES FOR A 1 MICROSECOND PULSE	TECHNOLOGY TIMEFRAME
VLSI/ASIC	10^{-6} to 10^{-8}	1980 to present
Low-noise transistors and FETs	10^{-7} to 10^{-8}	1980 to present
Digital ICs	10^{-6} to 10^{-3}	1970–1980s
High-power transistor	10^{-3} to 10^{-1}	1960's to present
Wire-wound resistors	10^{-1} to 10	current
Valves	10^{-1} to 10	1900–1960s

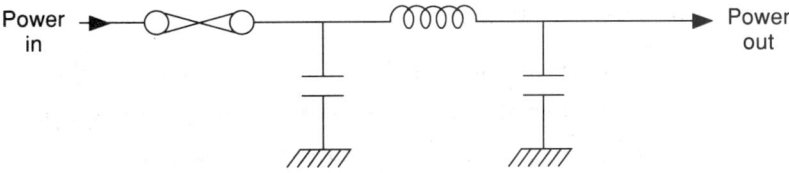

Figure 12.7   *The circuit of a single-phase surge filter.*

Power-Line Shunt Protection

The protection offered by shunt devices is limited and requires very good installation practices to be effective at all. The clamping can leave relatively high residual voltages.

Power-Line Filters

Power-line filters are installed in series with the equipment to be protected. Being similar in principle to power-line surge protectors, both common-mode and differential-mode protection is provided. These devices are often small self-contained plastic or metal boxes that rely directly on the ground from the power outlet. Typical power ratings are 1 to 15 amps, with surge ratings of around 5 kA.

Figure 12.8   *A three-phase surge filter. Photo courtesy of Critec.*

Line Conditioners

The main purpose of a line conditioner is to regulate the supply voltage to the protected equipment. A common version uses a transformer with a number of taps that can be connected as required to obtain the desired power output.

Communications Lines Protection

First level.   Simple shunt devices such as gas discharge tubes can provide some protection. The devices are slow-acting but cheap. They are usually found on the subscriber's termination point of the MDF at land-line telephone switches.

Second level.   A combination of series transorbers and shunt gas arresters can significantly improve the line protection. These are still relatively cheap but not completely reliable.

Third level.   Complex combinations of transorbers, gas arresters, and filters are used when maximum protection is required.

Working Voltages

Telephone lines are usually clamped at 200 volts. This is largely because of the ringing voltage, which can reach 130 volts. Other clamping levels are 7.5, 15 (for modems), 30 (RS232), 68, and 135 volts.

Coaxial Line Protection

Coaxial line protectors are available with clamping voltages from 90–1000 and can be provided with N, BNC, UHF, and IBM connectors.

Transient Ground Clamps

Where the local regulations provide for isolated grounding, an extra degree of security can be provided by placing earth clamps between the earth systems, which are normally open-circuit (and so preserve the grounding isolation) but which will break down under surge conditions to clamp the grounds together.

RFI Filters

Where the source of interference is a radio frequencies series, RF or HF filters can be placed in the lines.

Fiber-Optic Cables

These naturally provide good immunity against virtually all surges.

LIGHTNING PROTECTION

Lightning strikes the earth at a rate of about 1000 times per second. It kills five people per year in the UK, 14 in Japan, and 90 in the U.S. During the last 15 years it has been responsible for the loss of nine aircraft in the U.S. and has caused damage to aircraft ranging from minor pitting to punching 20-centimeter holes in thousands of others. Lightning most frequently strikes as a result of storms but has been recorded in clear, blue sky conditions. Damage is most prevalent when the soil is of low conductivity. Granite hills are particularly vulnerable locations. A radio antenna on top of a tower is especially at risk for anything from minor pitting to total vaporization.

For many centuries the most likely target for lightning was the village church, which was usually the tallest building in the town. Long after Franklin had clearly established that effective protection could be made available with the installation of a simple lightning rod, many churches refused to install the devices, claiming that lightning strikes were the will of God and nothing should be done to interfere. Many churches, like the one shown in Figure 12.9, were needlessly destroyed before this policy was abandoned. This spire carries a plaque reading "The spire of St. Bride's church, Fleet St. London. Removed in 1764 due to lightning and erected here, Park Place Estate in 1837." It now stands in an open paddock, only a few hundred meters from a tower jointly used for base stations of British Telecom and Racal-Vodaphone.

Lightning protection works by providing a path of low resistance for a lightning strike. For this reason, the lightning conductor must be the highest point on the tower and have a good path to the ground. That path is best provided by copper straps.

Since the days of Benjamin Franklin (who invented the lightning rod), some people have argued that the protection of a lightning rod comes from "discharging" the atmosphere around the tower and preventing strikes by preventing static build-up. This belief has been soundly disproved by every generation since Franklin, but it still persists. Even today some manufacturers claim to produce such devices. There is no evidence that pointed lightning rods work better than rounded ones.

Figure 12.9   *The remains of a London church, struck by lightning in 1764 and reassembled in a field near a modern base-station site.*

A very dangerous variant of the "discharge" type of lightning rod, which was popular a few decades ago, is the *radioactive* prong rod. In this instance a small capsule of radioactive material was embedded in the end of the prongs. The theory was that the radiation thus produced would ionize the air adjacent to it and so provide a low-resistance path for the discharge. Although the amount of radiation from such a rod was minimal most manufacturers have discontinued the production of such devices. Today, when the old rods are rapidly decaying to the point where the radioactive capsules are exposed (or may even fall to the ground), the hazard is greatest.

Lightning rods are the first line of defense. They should be placed to minimize potential interference with RF propagation and to maximize protection. No matter how often the warning is repeated, there is always somebody willing to tempt fate—as in the installation in Figure 12.10.

The "zone of protection" can be defined as the 90-degree cone around the antenna, as shown in Figure 12.11. The protected antennas should be inside the "cone" of protection described by the lightning rod.

Figure 12.10   *An omni mounted well above the lightning rod is tempting fate.*

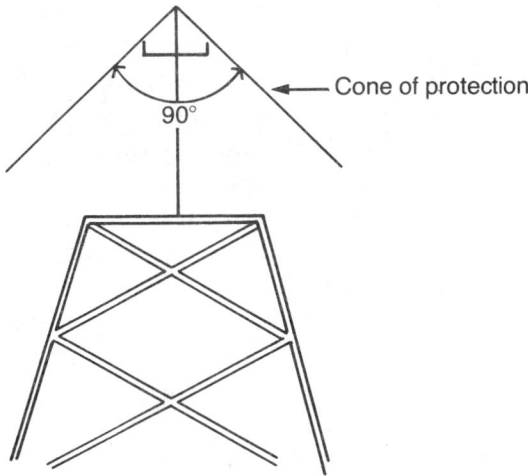

Figure 12.11   *Lightning rods can be envisioned as providing a "cone" of protection to an area beneath them.*

With the number of antennas on even a medium-sized common user site, it is often difficult to find a place to put a lightning rod where it will not cause significant pattern distortion. For this reason, many operators dispense with the lightning rods and rely on DC ground-potential antennas (antennas designed to withstand lightning discharges). This approach seems to work reasonably well.

Because the antenna feeders have large-diameter copper shields, they can make very attractive lightning paths. To reduce possible equipment damage, it is good practice to ground the feeders at the top and bottom of the tower, as well as at the entry point to the building structure.

Lightning protection is an essential consideration and all antenna-mounting structures should at least be fitted with a lightning rod that is well earthed (directly to a proper ground via a copper strap). The tower itself, as a minimum requirement, should be earthed at each leg to separate grounding rods. The rod should be tied together, with a buried bus-bar and the earth-ring should be tied to the building earth as shown in Figure 12.12.

THREE-PHASE POWER

The power supply may be single or three-phase. If a three-phase supply is used to power the base station, it can be used directly only if the rectifiers and air conditioners are three-phase units. Even in this case it is necessary to derive some single-phase supplies for auxiliary items such as the lights and power outlets.

Because three-phase power is not available at all locations, the decision to use three-phase equipment where appropriate means that the network will probably end up being a mix of three-phase and single-phase hardware. Consequently, spare rectifiers and air conditioners of both three-phase and single-phase-type must usually be kept.

Single-phase voltage feeds can be derived from three-phase supplies and the load shared between the phases. Figure 12.13 shows the relationship between the line and phase voltages of a three-phase supply from both star and delta transformers.

Providing a power inlet on the outside of the building so an emergency generator can be plugged in if required is a good idea. If an emergency generator is used, a suitable isolation switch should be provided at the switch board to enable the generator to be engaged and disengaged without danger to the operator or the equipment; a three-position switch that includes a neutral position will suffice.

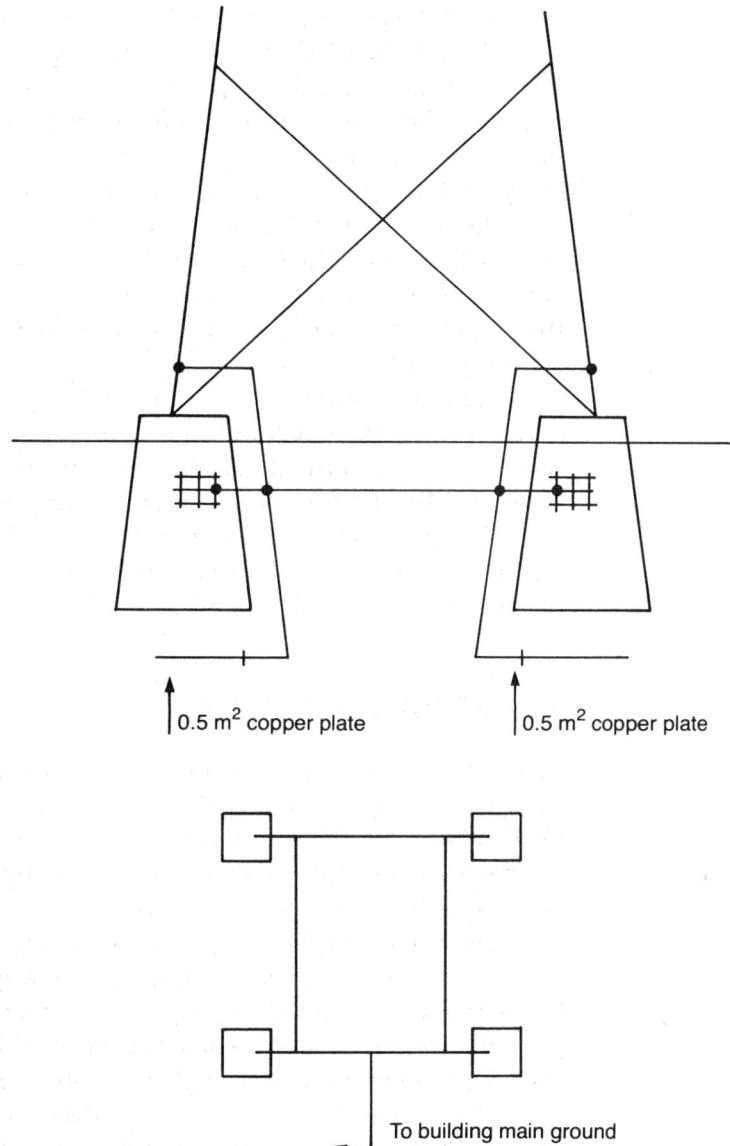

Figure 12.12   *A tower should be grounded at each leg and the grounding rods should be connected together.*

Power companies generally supply bulk power in a three-phase form because in this form transmission costs are lower than for a single supply. Consider the delta connected supply in Figure 12.13. If all the loads are equal, then the power supplied to the three loads = 3 ×

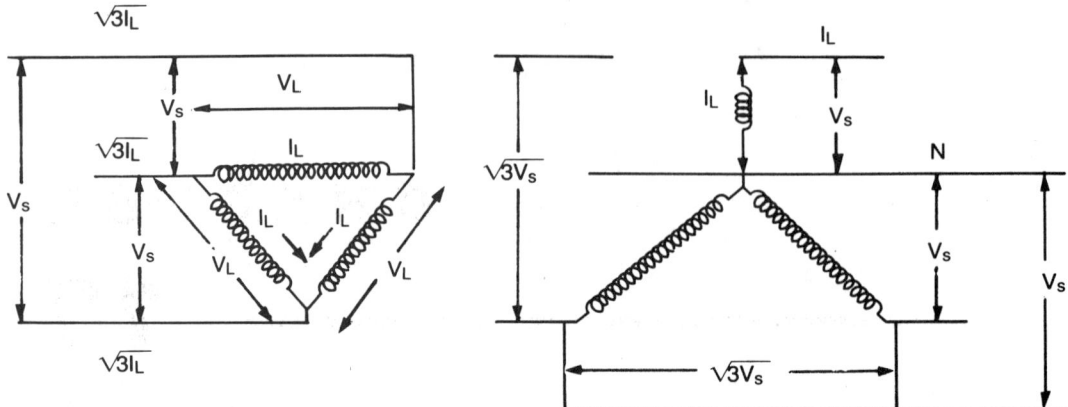

Figure 12.13   *Power grid three-phase supply configurations. Star- and Delta-connected main power transformers showing the voltages from line-to-line and line-to-neutral as a function of V_S (the single-phase voltage, which is usually 110/220/240 volts) and the current relationship for three balanced (equal) loads.*

$V_L \times I_L \times Pf$ (where Pf = power factor). The corresponding current carried by each of the three feeders is $\sqrt{3}I_L$.

If these loads were supplied by a single-phase line, the line current would be $3 \times I_L$, as shown in Figure 12.14. Hence, two conductors carrying $3I_L$ each would be necessary. The conductor size is a function of the current carried. If $I_L = 10$ amps, then the single-phase conductors would need to be AWG 6 or 4.13 mm. A 1-km feed length would require 237 kg of copper (two wires).

A three-phase conductor feeding the same load would need to carry $\sqrt{3} \times 20$ or 34 amps and would be AWG 10 or 2.59 mm. A 1-km feed length would require only 141 kg of copper.

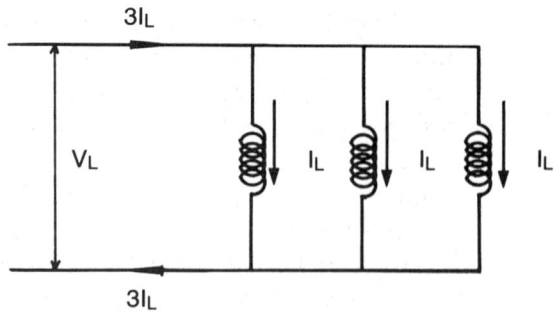

Figure 12.14   *The single-phase equivalent of Figure 12.13.*

 INSTALLATIONS

Most operators opt for turnkey installation, at least for their first system. A turnkey installation is one where the installer (usually the supplier) contracts to provide, design, and install a complete system and hand it over to the operator when it is fully functional (that is, ready to "turn the key" and start).

With turnkey installations, the operator must rely heavily on the supplier; although this may be necessary for new operators, the operator should take steps to avoid prolonged dependence. The disadvantage of being dependent on a supplier becomes obvious when a contract is prepared. An inexperienced operator must either write a very open-ended contract or employ a consultant. Both approaches have disadvantages, but the first is the most fraught with problems.

Suppliers usually dread an open-ended contract because they must specify their offers without any firm knowledge of what other suppliers might be specifying. A supplier who specifies a well-designed and complete network may lose the contract simply because another supplier cut corners to achieve the lowest bid price. A minimum design may seem more attractive to an inexperienced operator because omitting one base station can reduce the cost presented in a proposal by about $50,000 (including links, tower, and site works).

The disadvantages of open-ended contracts become apparent when the operator begins evaluation. Because the suppliers are not contracting for exactly the same thing it becomes very difficult to compare the offers. It takes a very experienced engineer several weeks to effectively compare two dissimilar proposals.

Relying on the supplier often runs smoothly through the design and installation phases but again becomes awkward during the commissioning and acceptance phases. At these stages it is necessary to certify that the work to date has been done adequately and in accordance with good practices and the terms of the contract.

Unless an independent appraisal of the work is available, the operator will find it very difficult to have faith in the acceptance. Even when a supplier goes to some length to ensure a fair acceptance test procedure is followed, the operator can never be completely sure of the value of that acceptance.

Depending on a supplier for ongoing expertise to keep the system operational also presents problems. When such expertise comes from the supplier, it is very expensive, and the operator does not have complete control over the availability and selection of expert staff. The alternative, using consultants, also has difficulties. With consultants, the operator also has little control over the availability and selection of expert staff. If the peak demands of the operator and the consultant coincide, the operator may not receive the highest-priority service. Therefore, any arrangement with consultants should include a retainer and a guaranteed response time.

Large consulting firms often assign different experts to different phases of the project. This division of labor causes a discontinuity in direction, and it may be that none of the consultants will have an adequate overview of the operator's system to be fully effective. It is wise to require that at least one specific consultant be available for the duration of the project.

Finally, the operator must assume that the consultants are competent because, almost by definition, the operator is not in a position to determine if a particular consultant is competent or not. For this reason the operator should acquire the expertise needed to design and run a network as early as possible.

TRAINING

Most systems offer a training program. These programs typically cost $500 per day per participant; the total cost for a start-up system would be about from $10,000 to $50,000. At this price, make sure to get good value for the money spent.

Begin by ascertaining that personnel have the educational background appropriate to the courses offered. The training courses usually assume a good technical knowledge of radio transmission for the RF courses, and good knowledge of switching equipment for the links and switch-training courses. The courses are usually detailed and complex and will soon leave behind those who do not have adequate technical background. Personnel must also have a good command of the instructional language (usually English). Insist too that training instructors be experts with a good command of the instructional language.

Courses should be provided in a timely manner; remember that a course undertaken and then not applied for six or more months will largely be forgotten. Course participants should be able to apply their new knowledge within two months of the course. The very fact that the knowledge will soon be needed increases retention.

By participating in the installation phase, employees can gain valuable insights into the functioning of the system. It is not unusual to specify in a turnkey contract that the operator's staff provide some of the labor for installation, thus ensuring active participation from the start. The areas of particular interest to an operator are the design phase (site selection), survey technique (RF path survey), site preparation, and, finally, installation and testing.

Because it will likely be necessary to constantly expand the network, some knowledge of what is involved in such expansions is also invaluable.

THE OPERATOR'S RESPONSIBILITY

When the operator accepts a turnkey system, payment becomes due and it is usually difficult to get the contractor to return for more than minor adjustments. Because the contractor's view of the project is somewhat different from the operator's, it is a good idea for the operator to be particularly alert during acceptance.

The operator's priorities are as follows:

- A good, efficient system with good coverage
- Ability to meet operational targets
- Ability to meet the market requirements at the least cost
- A competitive system
- Low maintenance costs
- Ability to expand efficiently and at minimum cost

A contractor's priorities, however, are somewhat different, namely, to

- Install the system on budget
- Meet the contract specification
- Perform as a credible contractor and win subsequent expansion contracts
- Meet time constraints

The order of these priorities is not always necessarily the same and some operations may have additional priorities, but it is easy to see that the objectives of the contractor and the operator are somewhat different.

For example, if a radio survey indicates that a particular area has marginal coverage and there is a possible (but not definite) need for an additional base, the operator, considering mainly the cost, may decide to take the risk and save money. The contractor, on the other hand, seeing the potential damage to the firm's reputation should poor coverage result, may decide that the doubtful base should be put in as a precaution. Alternatively, the contractor may ignore the problem in order to produce a lower quote. Whatever the contractor's decision, it will be based on different considerations than the operator's.

Contractors are often stressed by the demands of operators and the inability of manufacturers to supply on time. It is very easy, under such duress, to see the fine details such as labeling and documentation as relatively unimportant. The unwary operator who does not carefully check both the overall performance and the detail ultimately pays the price.

ACCEPTANCE TESTING

The operator should be responsible for acceptance testing because this is the one opportunity to ensure that all is well before paying for the system. At the end of this chapter are checklists that detail the items that must be checked before a base station is accepted into service. This checklist should take about one hour to complete.

Acceptance can be either absolute or conditional. If the base is inspected and found to have only a few minor shortcomings (for example, missing labels on equipment racks, some handbooks missing, and some spare parts not available), then it may be appropriate to issue a conditional acceptance (that is, the work is accepted subject to the shortcomings being cleared up within an agreed period—say one month).

But it is not appropriate to issue any kind of acceptance if major shortcomings are found. Examples would include the following:

1. The radio link to the controller is not functional.
2. There are no commissioning test sheets.
3. The installation is untidy.
4. Grounding straps are not provided.
5. The battery electrolyte levels are incorrect.

Because a poor standard of installation results in high maintenance costs over the life of the system, it is necessary to be very firm about acceptance procedures.

COMMISSIONING

As part of the acceptance, the accepting officer should be involved in the commissioning phase, usually in the last two days for a base station and in the last four to six weeks for a controller installation. This phase involves testing and aligning the system to ensure that everything operates within specification. The accepting officer should verify that all tests were properly done and recorded. The best way to do this is to be directly involved in the commissioning. Also, this phase is the most instructive part of the installation, and it is an opportunity for the operator's staff to become familiar with the equipment.

A physical check (using a vehicle and a number of pages) that coverage is adequate and that handoff occurs correctly should be done for each base station site. The coverage of the site should be confirmed manually (using a pager to see the limits of its range) or as a measured field strength. Serious discrepancies between actual coverage and predicted coverage may well point to some problems with the antenna, feeders, or system parameters. For this reason the operator should have access to a field-strength meter (using high-speed sampling). This is a lot cheaper than having the installer do those measurements, and it reduces the operator's dependence on the supplier.

MOVING AWAY FROM TURNKEY INSTALLATION

As the system evolves, the operator will likely gain confidence and be able to undertake a good deal of the work involved. Moving away from turnkey installation often results in large reductions in installation costs, and on that basis it should be at least considered by all operators.

Providing towers, huts (shelters), power, and clearing and preparing sites is work that can most readily be done first. The staff involved in this work should use the expertise available from the original turnkey project to become familiar with the requirements. The following sections discuss preparing the site for installation and establishing a staging area.

Site Preparation for Installation

It is essential to have a properly prepared site before installation can begin. A well-prepared site will be cleared and sealed in such a way as to ensure that a path to and from the site is free of dirt, dust, or loose particles. Other work that is likely to produce dust (for example, building extensions, preparing the site, and landscaping) should not take place simultaneously with installation. Power should be available on the site. The electrical grounding should be in place, connected, and tested. There should be a communications link back to the switch and, preferably, to other places as well. This link could be a telephone, a two-way radio, or even the engineering order wire on the microwave. The tower (support structure), antennas, and feeders should all be in place. Air-conditioning should be installed and operational, and all doors should be fitted and functional. The doors should have adequate security locks.

ACCEPTANCE-TEST SHEETS

The following acceptance-test checklists can be used by acceptance-test personnel for paging base sites.

Site (Name)   _____

Location   _____

Controller (Location)   _____Tel._____

Installed by   _____

Installation supervisor   _____Tel._____

Inspected by   _____

   ☐ on completion   ☐ work still in progress

Acceptance date   _____

Signed   _____

Conditional acceptance date   _____

Signed   _____

(Subject to rectification of items indicated on attached sheets)

Date in service   _____

POWER RECTIFIERS

		OK/NOT OK	COMMENTS
1.	Mounting and layout	_____	_____
2.	Cabling and terminations	_____	_____
3.	Alarm extension	_____	_____
4.	Designations	_____	_____
5.	Commissioning test results	_____	_____
6.	Handbooks	_____	_____
7.	Load sharing	_____	_____
8.	Safety signs	_____	_____
9.	DC distribution	_____	_____

BATTERIES

		OK/NOT OK	COMMENTS
1.	Fusing	_____	_____
2.	Battery spacing	_____	_____
3.	Electrolyte level	_____	_____
4.	Battery vents	_____	_____
5.	Battery lead burning/connections	_____	_____
6.	Battery cabling	_____	_____
7.	Hydrometer and thermometer	_____	_____
8.	Cell voltmeter and millivoltmeter	_____	_____
9.	Designations	_____	_____
10.	Drip trays	_____	_____
11.	Safety equipment	_____	_____
12.	Safety signs	_____	_____
13.	Water supply	_____	_____
14.	Fuse alarm extension	_____	_____
15.	Accessibility for testing	_____	_____
16.	Battery function and continuity (test)	_____	_____
17.	Floor loading of batteries within limits	_____	_____

EXTERNAL PLANT

		OK/NOT OK	COMMENTS
1.	Tower, mast, or pole	_____	_____
2.	Gantry	_____	_____
3.	Guys and anchors	_____	_____
4.	Lightning protection	_____	_____
5.	Tower, mast, or pole grounding	_____	_____
6.	Mains surge protection	_____	_____
7.	Tower lighting	_____	_____
8.	Corrosion protection	_____	_____
9.	Safety signs	_____	_____
10.	Site RF radiation records	_____	_____
11.	Equipment shelter	_____	_____
12.	Base security, fences, locks, gates, etc.	_____	_____
13.	Cable window and seals	_____	_____
14.	Grounding of equipment rooms	_____	_____
15.	Access to tower restricted	_____	_____
16.	Antennas correctly mounted	_____	_____

INTERNAL PLANT

		OK/NOT OK	COMMENTS
1.	Mounting and layout	_____	_____
2.	Cabling and terminations	_____	_____
3.	Alarms and telecontrol	_____	_____
4.	IDF cabling	_____	_____
5.	IDF labeling	_____	_____
6.	System performance	_____	_____
7.	Commissioning test results	_____	_____
8.	Station logs	_____	_____
9.	Designations	_____	_____
10.	Handbooks	_____	_____
11.	Drawings	_____	_____
12.	Test cables and extender cards	_____	_____
13.	Spares	_____	_____
14.	TRX plug crimps	_____	_____
15.	RF N-connectors	_____	_____
16.	Independent link to switch (control center) functional (telephone line or radio link)	_____	_____
17.	Lightning arresters on all incoming cables	_____	_____
18.	Door alarms functional	_____	_____
19.	Suitable fire extinguishers	_____	_____
20.	Redundant control channel functional	_____	_____

ITEMS TO BE CHECKED

Power Rectifiers

Mounting and layout

☐ Correct positioning of cabinets.

☐ All mains terminals covered.

☐ All cabinet components supplied, including tops and coverplates for unused positions.

☐ No cracked or non-working meters.

Cabling and terminations

☐ Cabling runs are satisfactory, cable ties used where necessary.

☐ Correct size of cable used for current to be carried for maximum size of installation.

☐ Cable crimps not loose.

☐ No undue mechanical stress by heavy cables on circuit breakers.

Alarm extension

☐ Correct settings and operation of all alarms provided on power cabinets (mains fail, float low, high volts, etc.)

Designations

☐ All circuit breakers labeled.

☐ Cabinets (if more than one) are numbered.

☐ Switch plates clearly marked.

☐ Modules are all numbered, if modular-type rectifiers.

Commissioning test results

☐ Should be provided by the installers and be on site.

Handbooks

☐ Should be left on site by the installers (and usually contain the test results).

Load sharing

☐ If power supply is modular, all modules should supply approximately the same current (but not necessarily equal). Turn the rectifiers on one at a time and note reconfigured load sharing and current limiting functionality.

Safety signs

☐ Mains hazard.

☐ –Ve 24 V ground (where applicable).

☐ Any others that are required by local regulations.

DC distribution

☐ Cabling from power cabinets to radio cabinets for correct current rating.

☐ Trays used where necessary to support the cable correctly between cabinets.

☐ Bus bars and battery feeders insulated and suitably protected against accidental short circuits.

Batteries and Distributions

Fusing

☐ Current rating of fuses supplied for battery capacity and current drains.

☐ If indicators supplied, these should be clearly visible.

☐ Spare fuses available.

Battery spacing

☐ Clearance over and around cells is sufficient for maintenance work (SG readings, etc).

Electrolyte level

☐ Correct in all cells.

Battery vents

☐ These can be frail and can be easily broken in installation. Check none are broken.

Battery lead burning/connections

☐ Each cell is correctly connected to the next via the lead V-connection. There should be no cracks or breaks capable of producing a high-resistance across the connection.

Battery cabling

☐ Cables correctly tied and supported on cable trays.

☐ Correct size cable used.

☐ No loose crimps.

Hydrometer and thermometer

☐ Correct size hydrometer is on site.

☐ Thermometer is present in each pilot cell.

Cell voltmeter and millivoltmeter

☐ Both on site.

☐ Spiked voltmeter 0–3 V to measure cell voltage.

☐ Millivoltmeter to measure the volt drop across the inter-cell connections.

Designations

☐ Each cell and battery is correctly labeled.

☐ Pilot cell is indicated.

☐ Fuses are labeled.

Drip trays

☐ Size and capacity are adequate to catch and contain one cell full of acid if a cell container cracks.

Safety equipment

☐ Check presence of equipment as required by company regulations (can include rubber gloves, rubber apron, face shield, first-aid kit, etc.)

Safety signs

☐ Acid precautionary warning signs posted.

Water supply

☐ Fresh, clean water and a small washbasin available.

Fuse alarm extension

☐ Alarm given when fuse open circuit or circuit breaker operated.

Accessibility for testing

☐ Batteries should be placed in such a position to allow cell replacement without impediment.

Battery function and continuity

☐ Gradually reduce the rectifier output voltage and note changeover to battery-powered operation. A voltage drop of not more than 2 volts should occur and current should remain the same. This checks battery and battery feeder continuity as well as the changeover mechanism. In systems not yet commissioned, the rectifier power should be turned off for this test. The battery voltage will initially drop rapidly and will then rise stabilizing about one volt above the minimum.

Floor loading of batteries within limits

☐ Battery loads (kg/m^2) should be confirmed as being within floor structural limits.

External Plant

Tower, mast, or pole

☐ State which, and give height. Check all ironwork is galvanized, no evidence of early rust, and all nuts and bolts are in place.

Gantry

☐ Feeders between structure and building are adequately supported by a gantry or tray.

Guys and anchors

☐ Check masts having multiple guys and concrete anchors. Guys should be examined for correct tightness and rusting, and concrete anchors for flaking and cracking.

Lightning protection

☐ Usually provided at top by antenna. Ensure lightning rods will not interfere with antenna pattern.

Tower, mast, or pole grounding

☐ Structure to be strapped to ground at ground level. Feeders to be grounded top and bottom of tower and at cable entry. Tower grounding connected to building ground.

Mains surge protection

☐ Usually provided on mains powerline into building.

Tower lighting

☐ Provided on structures near airfields, in accordance with local aviation regulations.

Corrosion protection

☐ All ground strap connections are sealed with anti-corrosion kit.

Safety signs

☐ Relevant safety signs are prominently displayed at foot of structure and on site fence.

Site RF radiation records

☐ When required, this record contains the maximum radiation from each antenna, and the safe working level of radiation on the tower top.

Equipment shelter

☐ The equipment shelter is properly and completely finished.

Base security, fences, locks, gates, etc.

☐ The site fencing and security are ensured.

Cable window and seals

☐ Cable window and seals are properly fitted to prevent water seepage.

Grounding of equipment rooms

☐ The equipment room is adequately grounded and the ground resistance less than 10 Ω. The test results should be available.

Access to tower restricted

☐ The tower is separately fenced and locked.

Antennas correctly mounted

☐ Antennas are correctly spaced and either vertical or at the correct level of downtilt.

Internal Plant

Mounting and layout

☐ Positioning of cabinets and supporting framework is as per design drawing.

☐ Blank panels, covers, etc., provided as required.

☐ Feeder supports provided above cabinets and up to the cable window.

☐ Frame racks firm and secure.

Cabling and terminations

☐ Neat distribution and positioning of all inter-cabinet cables; tied down at regular intervals.

☐ Cable plugs to be complete, no missing components.

☐ Plug labels correctly supplied and marked.

☐ Combiner-module and combiner-star connector tails are free and unstrained; N-connectors to be tight.

Alarms and telecontrol

☐ All alarms as specified are correctly returned to the mobile switch or monitoring center.

IDF cabling

☐ Neatness and tying of cables.

☐ Terminations correct for the type of termination applicable at IDF.

IDF labeling

☐ All circuits correctly labeled appropriate to the type of IDF system.

☐ Record book correctly made out.

System performance

☐ All channel modules within specification.

☐ All combiners within specification.

☐ Feeder-antenna return loss within specification.

☐ Base station controller, redundancy functional (where fitted).

☐ PCM or link system.

☐ System line-up levels on each channel or port correct.

☐ All alarms functional.

☐ Final call-through test on each channel module prior to cutover.

Commissioning test results

☐ To be left on site by installers for subsequent use by maintenance staff.

Station logs

☐ To be provided at cutover by installation staff for batteries, attendance, or other as locally required.

Designations

☐ Cabinet numbering.

☐ Module numbering.

☐ –Ve ground signs.

Handbooks

☐ To be left on site by installers for maintenance use.

Drawings

☐ Copy of floor layout and cabling records to be left on site with handbooks.

Test cables and extender cards

☐ Available where needed.

Spares

☐ Check any spare parts ordered are available before cutover.

TRX plug crimps

☐ Check crimping of wire to plug connection.

RF N-connectors

☐ If available, use N-connector gauge on all feeder connectors and tails. Some center pins may be out of specification and can cause damage to the socket.

Independent link to switch (control center) functional (telephone line or radio link)

☐ Check for proper functioning of voice link to switch or control room.

Lightning arresters on all incoming cables

☐ All cables entering and leaving the building should have suitable lightning arresters.

Door alarms functional

☐ Test all door alarms and confirm the proper working.

Suitable fire extinguishers

☐ Suitable non-conductive fire extinguishers are in place.

Redundant control channel functional

☐ Turn off the operational control channel and confirm the proper functioning of the changeover.

CHAPTER
14

 # BASE STATION MAINTENANCE

Most paging systems have good housekeeping software; problems such as VSWR out of range, TX failure, power and entry alarms, and signal path conditions are constantly monitored and are automatically reported to the maintenance-control center.

Maintenance usually consists of board or panel replacement, and this replacement is directed by the base-station controller or by the switch (that is, apart from reporting the fault, diagnostics suggest the maintenance procedure).

Routine maintenance is usually handled by one or two technicians who respond to a fault by replacing the suggested boards or panels. It should include approximately monthly visits to check on the condition of the batteries, air-conditioning, and site upkeep. In particular, the grass and foliage around the site should be neatly kept to avoid fire hazards.

Because pager bases are often on remote sites, careful attention should be paid to vandalism and attempted vandalism. Having a good security fence and some means of preventing intruders from climbing towers and structures is essential.

If this routine maintenance was all that was required, then base-station maintenance would indeed be easy. But there are other considerations. The rest of this chapter discusses these considerations in detail.

MAINTAINING QUALITY OF COVERAGE

A VSWR alarm and low TX power alarm will generally suffice to detect most RF failure or partial failure, but not all problems.

The following problems can cause reduced radiated power without detection:

- Water in the antenna
- Partial lightning damage to antenna/cable
- New (since installation) buildings, foliage growth, tower extensions, and other obstructions
- Damaged feeder
- Damaged, faulty, or waterlogged connectors

Catastrophic and severe failures will, of course, be detected by the base-station controller; less severe failures may not be detected. Most of these failures will result in reduced coverage, but they may escape detection by the alarm system. This is particularly true in the case of installations with high feeder losses.

The VSWR is measured by comparing the level of the forward RF signal to the reflected signal, and if measured from the TX output, only the forward component is accurately measured. If the fault is at, or near, the antenna, the reflected component has effectively traveled the distance to the antenna and back, and has hence suffered attenuation = 2 × feeder loss. This is how serious antenna problems can escape detection.

The VSWR is most easily measured by measuring the actual reflected power through a directional coupler. Once this value has been measured, the VSWR can be calculated as follows:

$$VSWR = \frac{1 + \sqrt{\dfrac{\text{Reflected Power}}{\text{Forward Power}}}}{1 - \sqrt{\dfrac{\text{Reflected Power}}{\text{Forward Power}}}}$$

An inline RF power meter can measure both of these components. A base-station antenna should have a VSWR of no greater than 1 to 1.7.

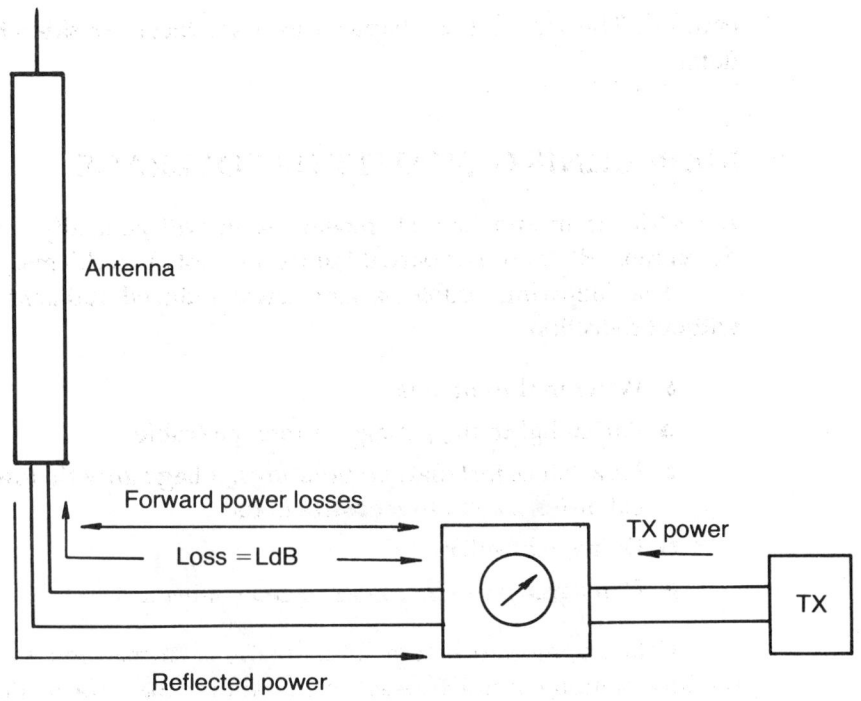

Figure 14.1 *VSWR measurement at the transmitter output terminals.*

If the reflected power is measured at the transmitter terminals, it is necessary to increase the value of the measured reflected power by the cable loss × 2 to convert the reading to the value at the antenna input. Figure 14.1 shows this VSWR measurement at the transmitter terminals.

About 10 percent of base-station sites can be seen by visual inspection to have at least one antenna mounted incorrectly. Omnis are often mounted so that with time they will slip to off-vertical. This will seriously reduce the range of the system. A simple visual inspection from the ground will often reveal the most serious cases like the one shown in Figure 14.2, where the two top-mounted omnis provide an easy visual reference that at least one of them is leaning. The visual inspection can be made more effective by using a pair of binoculars (about 8 × 20 will do). Such inspection should be a routine part of site maintenance. The alignment of the antennas should be checked by a rigger at least once a year, and as a matter of course during routine tower inspections.

Figure 14.2   *Off-vertical omnis, like those at the top of this tower, often account for poor coverage.*

CONFIRMING COVERAGE

It is important to track any degradation in the service area of a paging system. A reduction in coverage or an increase in lost calls can occur gradually and in such a way that it may become accepted by the users who do not detect it. The feedback between the person paged and the caller is often not good, and the caller may call more than once but not mention this to the called party. Thus a lack of complaints does not necessarily mean that all is well.

The most likely causes of lost calls are a reduction in the ERP of a transmitter, or a phasing or delay problem caused by some change in the links. ERP problems can arise from many faults, including water in the antenna or feeders, lightning damage, antennas that have moved off vertical, obstructions (including new buildings near either the transmitter or the receiver, and antennas mounted on the same tower as the transmitter that obscure the transmit path), a drop in transmitter power, frequency drift, and filters out of tune.

All multibase digital and analog systems have built-in delay lines, which can be adjusted to ensure that signals from the various

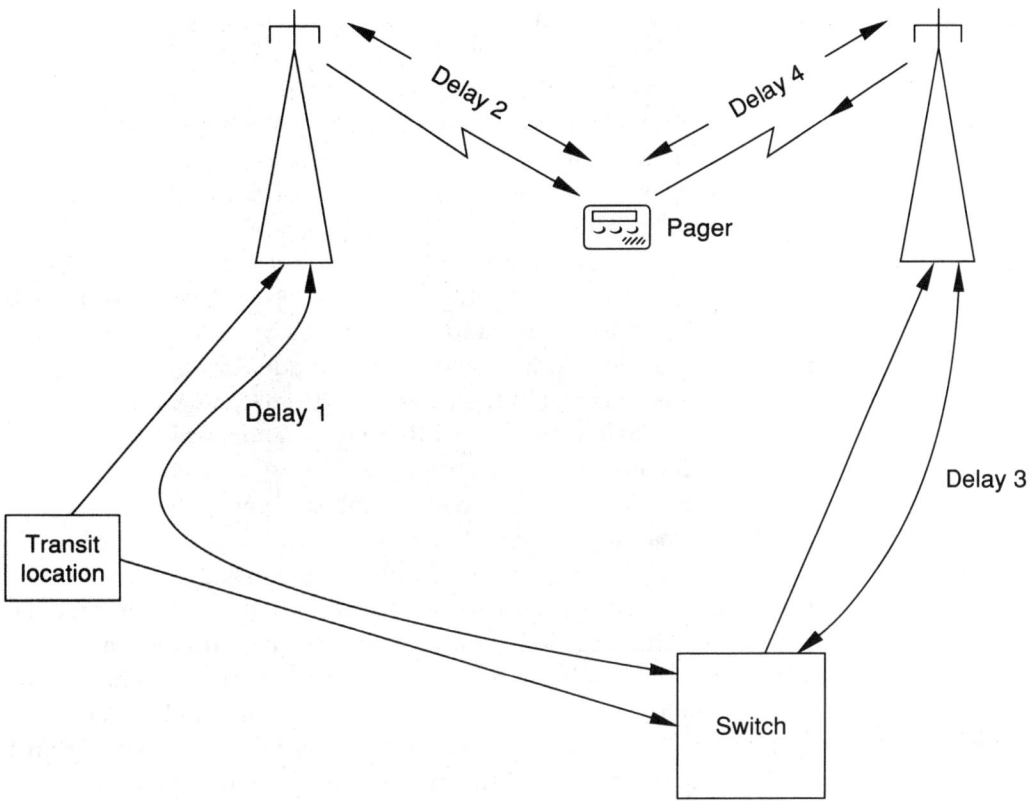

Figure 14.3   *The delays in the path of the paging system to the receiver are composed of a network part and an air part.*

transmitters in the network arrive in time synchronism. The delay, as seen in Figure 14.3, is composed of the air part and the land-line part.

From Figure 14.3 it can easily be appreciated that if there are any changes in the routing of the network path then the total delay may vary to the extent that data clashes occur at the receiver. It is unfortunate for paging-system operators that land-lines are frequently rerouted because of such things as replacement or upgrade of old cable, the establishment of new routes, or sometimes temporary rerouting during system maintenance.

With analog link systems, the situation is even more critical, because the phase of the incoming signal is also important and something as simple as transposing a pair of wires can lead to total cancellation of the audio signal at the receiver.

Whether or not the paging operator has control of the interconnecting links, these problems will occur sometime. This problem is recognized by the sudden appearance of a new "dead area" in which the RF field strength is still adequate.

Field-strength measurements will reveal a drop in ERP, where that is the problem, but they must be taken in areas where the results are unambiguous and the contribution of transmitters other than the one under examination is small. However, because delays and phasing can cause problems, a field-strength measurement in itself is not sufficient. Remember that to measure field strength, continuous measurements must be taken, so the transmitter being measured must be left triggered on for the duration of the measurement. Where this cannot be done, an allowance must be made for the significantly increased error margin in the field-strength reading. Under these circumstances, small decreases in ERP may go undetected.

A relatively simple but effective way to determine the service area in an absolute way is to use a number of pagers, which are separated by at least a wavelength and programmed on the same identity, to receive a cyclic page. For example, four message pagers, set with the same ID, may be paged every 30 seconds to receive a message. The regularity of the page will be affected by the queueing system, so to be sure that missed calls are recorded it will be necessary to have a fifth pager, which is connected to an external antenna and which can be assumed to receive a page even when there is insufficient signal to activate the other pagers. This master pager, then becomes the reference that a call was sent. The master may still fail to receive calls even in high field strength if delay problems exist, so a field-strength indicator of some kind is essential.

AIR-CONDITIONING

The integrity and proper functioning of the air-conditioning is most important to ensure reliability of the base-station equipment. It is well-established that temperature cycles contribute significantly to the fault rates of electronic equipment. It is a good idea to provide dual air conditioners so one unit is available if one should fail.

The trade-off of having the air conditioners out of service during power failures and the cost of providing emergency power should be considered. Equipment that has a low thermal operating range may not be able to function for any significant period without air-conditioning but some more robust equipment can. The fact that some equipment can operate at high temperatures, however, does not mean that it is immune to the deleterious effects of temperature cycling.

Routine maintenance consists essentially of changing the filters at regular intervals. These intervals are largely determined by the airborne particle concentrations in the area and can vary from weeks to months.

Maintenance and operations staff in warm climates should also watch for the formation of condensation, which can be indicative of excessive ingress of warm moisture-laden air. This can be readily rectified by blocking the air leaks.

However the most likely source of moisture remains insufficient waterproofing.

Condensation on the outside of the building indicates excessive heat loss through the shelter and can be rectified by improved insulation.

Failure to rectify these problems can lead to base-station failure due to moisture on CMOS boards or corrosion of boards and connectors. Also inefficiently operating air conditioners will consume excessive power.

A humidity meter should be placed in each base station and the levels recorded periodically. In general the humidity should be kept in the range of 35–65 percent.

POWER AMPLIFIERS

Without a doubt, power amplifiers (PAs) require the most consistent maintenance attention. Although suppliers' published figures for MTBF vary considerably, the probable MTBF in the real life of a PA is about one or two years. Published figures are invariably based on different measurement criteria of MTBF, which reflects sadly on efforts to standardize MTBF measurement.

The least reliable part of the transmitter is the RF power stage. Because its loss puts the base off the air, it is almost always recommended that redundancy be provided. For the same reason, when taken out of service, it should be attended to quickly.

CUSTOMER COMPLAINTS

Some authorities rely on customer complaints to inform them of ERP (Effective Radiated Power) problems of reduced coverage. This detection system, apart from causing poor public relations, will not work if there is significant base-station overlap and other bases take over from the faulty one, or if the fault is gradual (an example, water accumulation) and subscribers come to accept the "new" coverage as normal.

To make matters worse, many customer complaints are ill-informed. "The service doesn't work at my house" may mean "It doesn't work in my underground garage" but, of course, the service is fine outside the garage.

There can be a number of reasons why a customer reports less than satisfactory service.

- The pager is faulty.
- The pager battery is exhausted or providing inadequate power.
- The customer is outside the normal coverage area.
- A base-station fault is evident.
- The customer attempted calls during the down time of the system/base.
- The system is congested.
- The customer likes to complain.

At this point there is a need for clear communication between the person recording the fault, the mobile maintenance staff, and the base-station and switching staff, to avoid duplication.

This is an easy way to proceed:

1. The fault recorder has a coverage map, with good street-level accuracy, and can ascertain whether the complaint originates from within the normal service area. The fault recorder should ask sufficient questions to ensure that the customer is using appropriate operating procedures.

2. If the complaint relates to usage outside the service area, the customer should be advised accordingly.

3. If the complaint is inside the service area, then an appointment to check the pager should be arranged with the mobile maintenance staff. At the service center it may be possible to determine absolutely whether the fault is pager or not, but, due to lack of adequate test equipment, it may also be that trying a replacement pager is the only solution.

4. Unless the pager has been clearly implicated, the fault report should now be passed on as a possible base-station fault. The

fault could also have been a temporary controller or base-station problem (for example, outage due to maintenance). Because of this, these reports should be handled centrally and forwarded to the controller or base-station staff only when a clear pattern has emerged.

5. Analysis of consistent fault reports in areas thought to be well-covered may reveal local "dead spots." This information is most useful to system planners and designers, so it is important that reports are carefully processed and the results forwarded to the relevant personnel. It may also reveal a timing problem.

LINE-UP LEVELS

After replacing or repairing any MUX equipment, it is essential to check all affected line-up levels. Line-up levels and deviation should be routinely confirmed every six to nine months.

SITE AUDIO TEST LOOPS

Some suppliers include a site audio test loop to distinguish remotely between noisy channels that are caused by the switch to base link from those that are caused by the base RF equipment. This test tells staff located at the switch if the fault is a link or a base-station RF problem. Because each of these problems is likely to be serviced by different personnel, a good deal of diagnostic time can be saved by early identification.

INTERACTION WITH THE CONTROLLER

Base-station maintenance requires close cooperation with the technical people in charge of the controller, because most of the alarm information is available from them. It is important that the lines of communication between the controller technician and the RF technician are good.

A great deal of time can be saved by having an accurate diagnosis of the problem before staff are dispatched, particularly to a remote base. There are many diagnostics available at the controller that can be used to determine the nature of a fault.

Since any service affecting alarms will prompt a response from the controller staff, they should be notified before any channels are taken out of service.

In particular, it is most important to diagnose whether the fault is from the controller, the transmission link, the site controller, or the RF. Some manufacturers have remote-alarm monitoring equipment that can be located in the RF repair center. This equipment can reduce the dependence on switching staff to diagnose base-station faults.

System designers can assist by co-locating a switch with a base station so that the switch operators become more familiar with RF hardware. There is a tendency in paging (as in all other enterprises) to form the "us and them syndrome" (that is, the controller operator always tends to assume that the fault is in the domain of the RF staff and vice versa). (Of course, co-siting should be done only if it does not compromise the system design, but it is generally possible to achieve such a design.)

The controller can provide details on service affecting conditions such as channels out due to interference, low power, data errors on the link, timing and synchronization problems, as well as faults detected while routinely polling the test mobile.

SITE-LOG BOOKS

All sites should have locally maintained log books. In these log books are recorded all visits (entries) to the site and the purpose of each visit. Each entry should cause an alarm at the switch that must be "canceled" by the person entering the base station. The entry is recorded by both the switch attendant and the person at the base station. Table 14.1 shows a typical log book.

Table 14.1   *Typical entries in a log book*

DATE	ARRIVAL TIME	NAME	DESIGNA-TION	DEPART-MENT	PURPOSE OF VISIT	DEPARTURE TIME	SIG
09.10.89	08.30	S. Davey	Technician	Maint.	Replace PA	10.30	...
09.10.89	13.00	N. Kelly	Cleaner	Cleaning	Polish floors	14.00	...

CALL-OUT PROCEDURES

After hours, it is usual to have only a skeleton staff available. The staff may amount to only one person directly on call at the switch or even one person on call-out via alarm rerouting. In either case, it is unlikely that the person on call-out will be an overall system expert, so procedures that determine the priority of call-out must be established.

In any large city there may be considerable overlap in base-station coverage, so that the outage of any one station, particularly after hours, would not be service affecting. This means it is not noticed by customers. In even larger systems, two or three non-adjacent bases may be out of service without seriously affecting service.

Rules must be drawn up to enable the call-out staff to decide on a course of action. For example, consider a medium-sized city with 6 base stations. After discussions between switching RF and transmission staff, the following rules may be decided upon:

1. In normal working hours up to 4:30 P. M., any base-station outage is treated as urgent and must be attended to immediately.

2. Between 4:30 P. M. and 7:00 A. M., and from 4:30 P. M. Friday to 7:00 A. M. Monday, any two base stations may be out of service without call-out.

3. All urgent alarms are to be attended to by the switch operator (either an on-site operator or one with a remote terminal if provided). This operator will decide on any subsequent call-out procedures.

4. In both the switching and radio areas, at least three personnel, with a predetermined call-out priority, will be available. If called, the first party to answer will attend to the fault and determine what other call-outs are needed.

5. Each of the call-out personnel will be supplied with a pager and mobile telephone.

6. Special call-out procedures will be defined from time to time for special holiday periods.

EQUIPMENT

To service the base stations, it is necessary to have at least one dedicated vehicle and two technical staff. A station wagon or a small van is a suitable vehicle. The vehicle should be permanently outfitted with the necessary test equipment and tools.

For most installations, base-station servicing will be ongoing, on a daily basis. Spare cards and parts will normally be stored at a central area. Maintenance staff need frequent access to this storeroom, so a central location is most important.

The following is the minimum equipment necessary:

1. Two VSWR meters to 50 watts.

2. One general purpose mobile test set incorporating at least the following to 1 GHz:

   a. RF signal generator

   b. RF level measurement

   c. RF deviation measurement

   d. RF frequency measurement (accurate to $\pm$ 1 kHz)

   e. Audio modulation FM and AM (for analog systems)

   f. Preferably a spectrum analyzer

   g. SINAD measurement

3. Three digital voltmeters.

4. One spectrum analyzer to 1-GHz RF and 10-Hz IF resolution.

5. Specific equipment for servicing the particular manufacturers' bases (for example, PCBs, PCs, special cables and plugs). Note that most special cables and plugs should be stored at each base station.

Because the time to travel to and from a base station is often significant, the maintenance van should carry a good range of tools and miscellaneous parts (such as connectors, transition connectors, cables, and lugs). It is not usual to leave test equipment permanently on site at a base station.

The maintenance staff should also have some independent means of communicating with the controller, such as via the PSTN cellular phone or a land mobile. This is important in the event of link loss or complete system failure.

QUALITY AND CALIBRATION OF TEST EQUIPMENT

Paging network equipment represents state-of-the-art hardware that is beginning to approach the theoretical limits of performance at room temperature. For this reason, it is essential that only high-quality and well-calibrated equipment be used when servicing or adjusting the hardware. The old service monitor of the early 1980s is probably only accurate to ± 5 or 6 dB, and even most modern ones are struggling to achieve ± 3 dB.

The equipment used to service base-station equipment should be accurate to at least ± 2 dB. This rules out most old equipment and even new equipment that has not been calibrated in the past 12 months. Errors of frequency, signal level, modulation level, or VSWR can lead to poor system performance. Additional sources of error include poor-quality connecting cables, connectors, and the use of multiple adapters.

TEST SETS

General purpose test sets are available from a number of manufacturers. A good test set should include most of the following:

- SINAD (Signal to Noise And Distortion) measurement with the appropriate weighting
- SINAD measurement of out-of-band test tones
- Simulation of signaling tones
- A simple spectrum analyzer
- A frequency counter (to 1 GHz)
- A signal generator (to 1 GHz)
- Deviation measurement
- RF power measurement
- RF millivoltmeter
- DTMF decoder
- FSK decoder

A good-quality spectrum analyzer is necessary to adequately assess the transmitter performance. The analyzers normally found on service monitors do not have adequate resolution for this purpose and cannot be used. The spectrum analyzer should have a tracking capability so that it can be used to sweep filter and cavity assemblies. The analyzer needs an IF filter resolution of 10 Hz and should be able to display spurious and harmonic generation up to 4 GHz (accurate measurement above 1 GHz is not necessary).

QUANTIFYING COVERAGE PROBLEMS

Good baseline records of the original coverage must be available in order to determine if a problem has occurred since installation. There are two practical forms of this information:

- Detailed, repeatable survey results
- Point-to-point readings taken from the base stations to *fixed* antennas located at strategic points

If the first method is used, a survey vehicle and equipment will be needed for maintenance (for more information, see Chapter 5, "Radio Survey").

Transmit antennas can be checked simply by measuring field strength.

SPARE PARTS

Initially, suppliers should recommend which spare parts to store. Be a little cautious, however, that the supplier will sometimes reduce the number of recommended spare parts in order to produce a lower bid price. Of course, once the system has been purchased, it is easy to recommend a few more spare parts (the same thing sometimes happens with training and test equipment).

By far the most unreliable component is the RF channel equipment; adequate spare parts should be kept to allow for a one-year requirement.

Be sure to test all spare boards soon after delivery by placing them into service. A non-functional spare is of no use and a significant number of boards are delivered from the manufacturer as DOAs (Dead on Arrivals).

At least three spare antennas of each type should be on hand. These often have fairly long-lead times on delivery and are liable to fail in considerable number during the electrical storm season.

Disposable items, such as fuses and light bulbs, should be purchased in quantities adequate for two years. The frequency of failure, coupled with the cost of processing orders, makes small holdings economically unattractive. Good quantities of spare connectors should also be stored. It is false economy to save on spares, and adequate quantities should be held to ensure that they will be available when needed.

CHAPTER
15

 # BILLING SYSTEMS

The billing system is an important but independent part of a paging system. Most paging system suppliers do not offer a billing system as part of their product line and recommend that this be purchased from an independent supplier. Wireline operators will almost certainly already have a billing system for their other operations, and this can be adapted. Some operators incorporate the billing so that the customer receives the paging bill in the same format as if it were an additional wireline telephone; others prefer to bill separately.

In principle the concept of billing is simple. The switch records all the call details on a billing tape, the tape is later read by another computer that analyzes that tape, and from the called number and the length of time that the call was held up, the charges to the calling party are determined and billed.

In practice getting a billing system fully operational and effective is very time-consuming and requires the understanding on the part of the operator that just because it is all computerized does not mean that it is infallible. In fact if a newly installed billing system can compute the charges due so that 99 percent of the records are correctly processed then that would be considered rather good.

Billing systems have human interfaces, and data entries include the tariff table, the charges on a basis of the called number, the charging algorithm. This interface and the software itself is where problems arise. It is generally considered that computer code that has not more than one line of faulty code per 1000 lines is fully debugged. This one line is the gremlin in the system, and it usually surfaces only occasionally and usually in the most mysterious way. You can expect that the price of accurate billing is eternal vigilance.

The billing computer is physically separate from the switch; frequently, the only link is a tape. The raw billing information is collected by the switch but is not processed. The tape is then removed and sent to be read and analyzed by the billing computer.

This billing computer can be a minicomputer or a mainframe computer that has its own software to interpret the billing tape and then produce the subscriber's bill and the operator's ledger. Because various manufacturers have different proprietary operating systems, the billing (DAS) tape format is different for each equipment type. If the billing system provider has not previously worked with a particular switch, some software must be written to enable the tape to be read.

Charges vary widely from country to country and vary somewhat from operator to operator within the same country.

Charging can involve many parameters. These are some of the main parameters:

- Cost of local calls (incoming)
- Cost of airtime
- Call rate variation with time of day
- Charges for long-distance calls and wide area
- Charges may vary with call length (for example, the first frame may be charged at a higher rate than subsequent frames)

The following requirements should be built into the software:

- Frequency of billing (monthly, two-monthly, and so on)
- Charges applicable to roamers
- Billing format (what the bill looks like)
- Currency (this can present problems in countries where the exchange rate is thousands of units to the dollar so that existing billing formats may not be able to accommodate a sizable bill)

■ Language (sometimes the bill may be required in languages other than English)

Real-time (or hot) billing requires that the billing computer can access the switch in real time and produce a bill on demand. The advantage of this is particularly evident in the case of roamers and rental units, where an instant bill is needed. To keep the demands on the processor within practical limitations, only some of the customers are flagged for real-time billing. Typically, a real-time billing package may allow up to 1000 customers at a time to be marked for real-time access. Such a system could use an RS232 link between the billing computer and the switch. Figure 15.1 illustrates this.

Billing systems can perform billing and ledger functions only, or they can perform validation and produce Management Information Systems (MIS). There are considerable advantages in the more capable systems, but they are also more costly.

Billing systems that can perform validation allow customer data to be entered manually only once into the billing system. If validation is not available, the dual entries of customer data into the switch and to the billing system are sure to leave considerable room for false and missed entries. Once the system is bigger than a few thousand customers, it is almost impractical to use two parallel data entry systems.

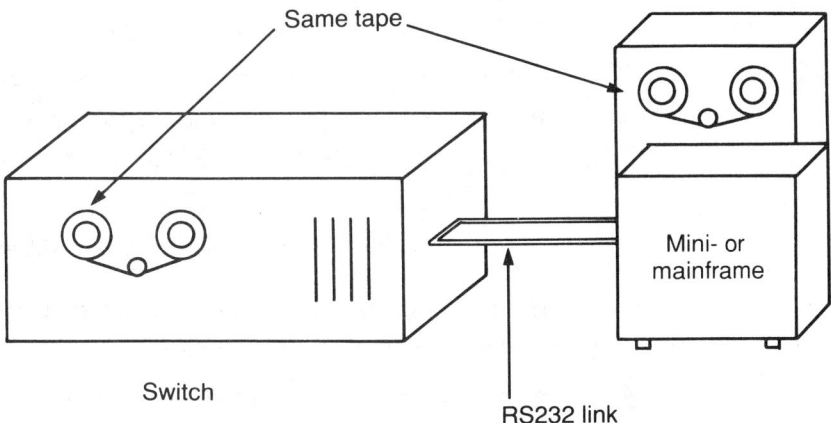

Same tape

Mini- or mainframe

Switch

RS232 link

Figure 15.1 *The link between the switch and the billing computer for real-time billing may use an RS232 link. The bulk of the billing information, however, is transferred via tape.*

THE DANGERS OF CONTRACTING
FOR A BILLING SYSTEM

Every operator must at some time obtain the services of a billing service and regardless of whether an in-house system, or outside billing service is chosen the operator will be dependent rather heavily on ongoing support for the billing software.

In one case that the author has been involved with, the billing supplier (for a cellular system) was somewhat less than reliable, and as soon as the payments had been made for the in-house system, the support dried up completely. A few months after the system was installed a severe power failure caused a number of the files to be corrupted, which caused the software to fail. This problem was never rectified by the supplier, and the billing system was never again functional. As a result the operator was unable to bill his customers. As a precaution it was necessary to cease to connect new customers until the problem could be resolved. In the meantime the lost revenue amounted to around $100,000 per month. A quick check of the contract revealed that the operator's recourse was to demand that the software be made functional (under the terms of the warranty). This was done but still no support was forthcoming. The final option available was to terminate the contract. Losses incurred as a result of the incompetence and/or bad faith on the part of the billing system provider were specifically excluded, as claimable liabilities under the conditions of the contract.

So in essence the billing supplier only guarantees to rectify the hardware/software. If however this is not honored the only recourse is to go to someone else. Doesn't really sound fair does it?

In subsequent dealings the author learned that all billing-system providers similarly require an indemnity against non-performance of their wares. Although this approach is understandable it does leave the operator very vulnerable to unscrupulous billing system providers. There is not a great deal of risk with most of the more reputable suppliers, but it does pay to check the reputation of the people you deal with very carefully.

FLEXIBILITY OF TARIFFS

Tariff structures are often quite complex and may have built-in day-, time- and distance-dependent rate structures. These rates are not fixed forever, and so it is important that the billing system offers an easy means of altering tariffs as rate changes occur.

ACCOUNT SETTLEMENT

The billing system must be able to produce a monthly statement of accounts due to the systems interconnected to the paging system. It may be that the settlement with various operators is structured quite differently and a good deal of flexibility is required for this part of the program.

BILLING CYCLES

The operator will have to decide on a billing cycle that maximizes revenue collection without unduly loading the billing system. Typically, billing cycles range from one month to three months.

When the system is large it may be necessary to have continuous billing cycles in order to avoid unmanageable volumes of paper, which would occur if all bills were sent on the same day.

ITEMIZED ACCOUNTS

These are readily available from the billing system and will give the customer details of each call, its duration, time of call (and sometimes the called number), and charges. Being detailed, these accounts can be bulky and it is not unusual that they are offered to the customer at an additional charge (usually a few dollars per bill).

REMOTE BILLING

The billing system will normally have the capacity to support remote billing terminals. Remote terminal access can be of use for customer-service staff who handle billing enquiries as well as for the use of remotely based billing staff who may be handling their own local areas.

FOLLOW-UP AND ACCOUNT MANAGEMENT

Not all bills are paid on time and delinquent bills need to be chased. A billing system can do this automatically and flag high-dollar bills for special attention. Automatic devalidation for payment default can be built in.

A series of reminder notices (also known as *dunning notices*) are sent to slow-paying customers to coax payment. These notices can be automatically generated by the billing computer. Usually a period of one to two months late payment is permitted before the pager is devalidated.

As the cost of carrying bad debts can be high, many operators charge interest on late payments. In the interest of good customer relations the interest rates should be reasonable and in line with the charges of other credit organizations.

BILL PREPARATION AND LETTER STUFFING

The billing system will normally prepare the customer's bill in a form ready for mailing. However simple systems may still require manual folding and insertion into an envelope. Additional hardware is available to automate all of these functions.

VALIDATION

Usually the operator will initially validate subscribers directly through the controller. Apart from the duplication involved in entering the data separately into the billing system, experience shows that the coordination between the two systems is often a problem. Billing systems can be structured so that an entry/cancellation on the billing system automatically updates the switch records.

TRACKING SALES AND INVENTORY

The billing system can be integrated into the sales network so that it can track the sales of pagers and keep an inventory of stock. In particular it can flag stock re-orders as well as especially fast- or slow-moving stock.

ON-LINE ENQUIRIES

The billing system should have on-line access to the billing records for the last two to three months. A readily accessible operator position should be available so that queries and complaints can be promptly dealt with. It pays to remember that even though the billing system is computerized it is far from infallible and that many of the discrepancies brought up by the customer will be legitimate. For those customers who make a habit of complaining some billing systems also have on-line access to previous customer complaint records. Typically you can expect around 1–3 percent of bills to result in customer queries. Disputes over charges can be damaging to customer relations and the more promptly and efficiently they can be handled the better.

BILLING SERVICE

The billing system should contain the following customer information:

- Age
- Occupation
- Sex
- Business/work address
- Average revenue
- Areas generating most revenue
- Revenue as a function of time of day (or week)

This information tells a good deal about the types of customers that are connected, but it can never yield the profile of the mythical "typical customer."

A marketing manager can use this information effectively. If it is used intelligently, this information can give a company a competitive edge. The goal, of course, is to get maximum return for minimum investment. In a mature network, achieving this goal means identifying and exploiting unused capacity and using existing capacity more efficiently.

Customers who have high calling rates (and hence generate high revenue) can be identified and marked for special attention to ensure that they are retained. To gain customer loyalty, perhaps package deals that offer reduced charges for those users can be arranged. The system could alert customers whose calling rates might entitle them to a better deal if they choose a package. A small drop in revenue is more than compensated for by gaining long-term, satisfied customers.

Incentive packages are often aimed at high-usage customers who, although they generate the most revenue, also use the system most and have a higher system cost per customer. In a fixed network, a customer who is connected but has a low usage rate costs the provider just as much as a high-revenue customer. Such customers are therefore undesirable to the operator. In paging systems, however, a connected customer who pays the monthly fee and doesn't use the system costs only the price of the billing services. Therefore, a low-usage-rate customer normally generates the best return/cost ratio to the network operator.

Sensitivity to pricing can be gauged by correlating new customers with price variations. Where the option exists (that is, it is not government controlled), this information could be very useful in determining the revenue mix from monthly access fees versus call rates. It may be found that particular locations are not generating their share of revenue and that a sales promotion in those areas is warranted. Or it may be determined that an area is intrinsically one of low-demand. Knowing why demands are low (or high) can be extremely valuable in future expansion plans. But even if the cause cannot be identified, some generalizations on the structure of the areas with abnormal demand can help to focus expansion.

Information from widely disparate sources reveals at least an order-of-magnitude variation in customer sign-up rates from similar sized cities with similar per capita incomes in different parts of the world. Within the same country, there are still large variations in what are otherwise similar markets. The best information, therefore, is obtained from the operator's own records; the marketing manager who realizes this is the one who will succeed.

CHAPTER

16

 INTERCONNECTION

Paging systems must be fully interconnected to the land-line facilities to be effective. The landline facilities have usually been in place for decades on a monopoly basis, and new competitors are not usually welcomed. This is a very difficult and often "emotional" part of the business. Generally there is no problem where the operator is also the wireline carrier, but in most cases the paging operator will be a new company that has to interconnect with the wireline facilities. A major hurdle placed in front of a new operator is to get over the legal and bureaucratic barriers that will be put up by the wireline carrier.

It must be understood that the wireline operator usually genuinely believes that he has a "right" to the monopoly that has been enjoyed over the years and that at the very least some compensation for the loss of that status is in order. Because of this, interconnect agreements that strongly favor the landline operator are common.

As the operator is not a subscriber, the inter-operator charges that apply should be cost based. That is, the compensation should be based on the cost of the facilities used by the parties plus a fair return on the investment.

Determining the value of the cost of interconnection is difficult. First it is necessary to define the point of interconnection. Ideally, the two operators would meet at a half-way point. This is illustrated in Figure 16.1.

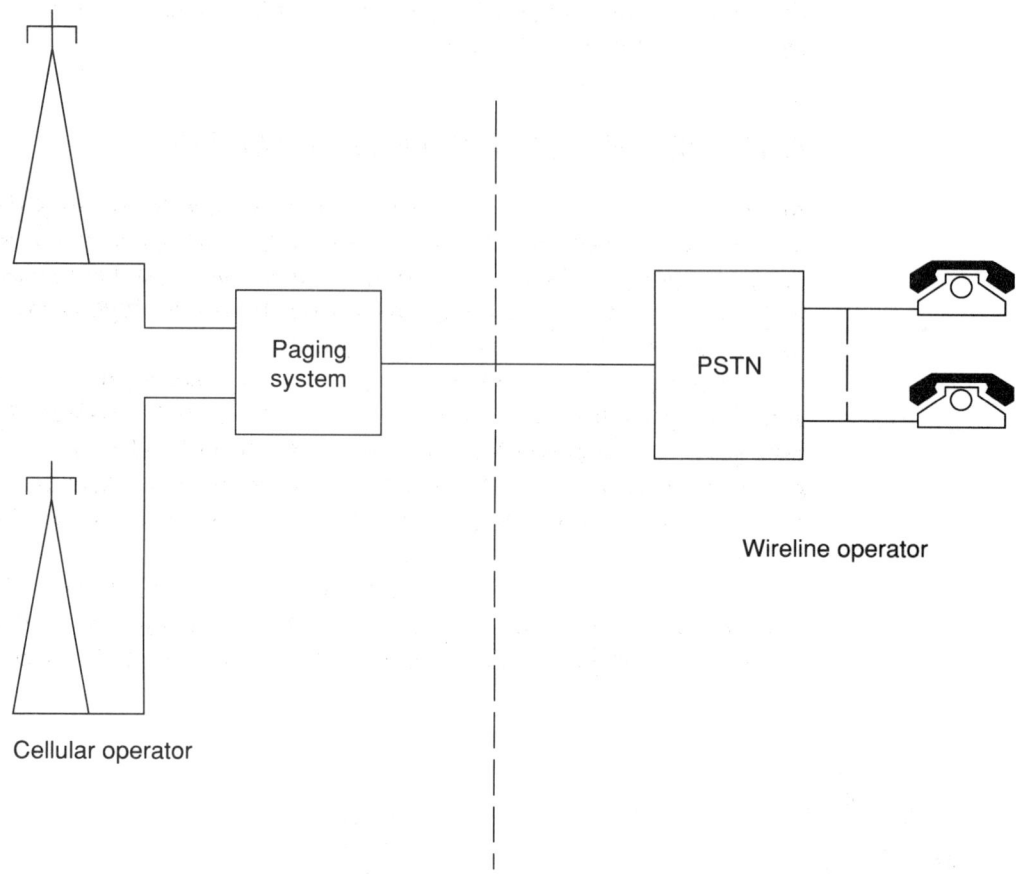

Figure 16.1   *The ideal point of interconnection is at the half-way point between the carriers.*

In most cases, the link between the paging system and the PSTN will belong to the PSTN operator, and it will therefore attract a monthly rental charge at a rate that would have been established for other carriers or lessees.

The most difficult part is to determine what is a fair and reasonable rate for traffic flows between the carriers. Immediately we can rule out using the subscribers call charges as these are the retail cost of a call. However this will set an upper limit for the fair cost of traffic.

Some carriers use a revenue-based settlement, but this is not accepted as a fair basis by the FCC. Most interconnect agreements will require that each operator be responsible for all calls generated in their own systems and also the cost of revenue *collection*, which must include a provision for bad debts. A call involves a PSTN caller using the land-line network and the paging subscriber using cost recovery should allow for this collection cost.

COST PER MINUTE OF TRAFFIC FLOWS

Many schemes have been proposed to allow the determination of the fair charge for interconnect. Those based on the cost of a switch inlet or outlet ignore the fact the first point of interconnect does not reflect the whole cost of carrying a call. The actual call path of a typical PSTN to paging subscriber can be seen in Figure 16.2.

In general, the fairest way will be to base costs on minutes of actual usage as this is very easy to measure and is easily verifiable by both parties. This differs from revenue, which will often be on a fixed cost per call, or for toll calls on a 1 + 1 per-minute basis (that is, on answer a one-minute charge is already levied regardless of the actual talk time).

There are a number of ways that the cost of an average call could be determined. The simplest is on the the average-cost-per subscriber basis. For example, a 10,000-line paging system may have cost

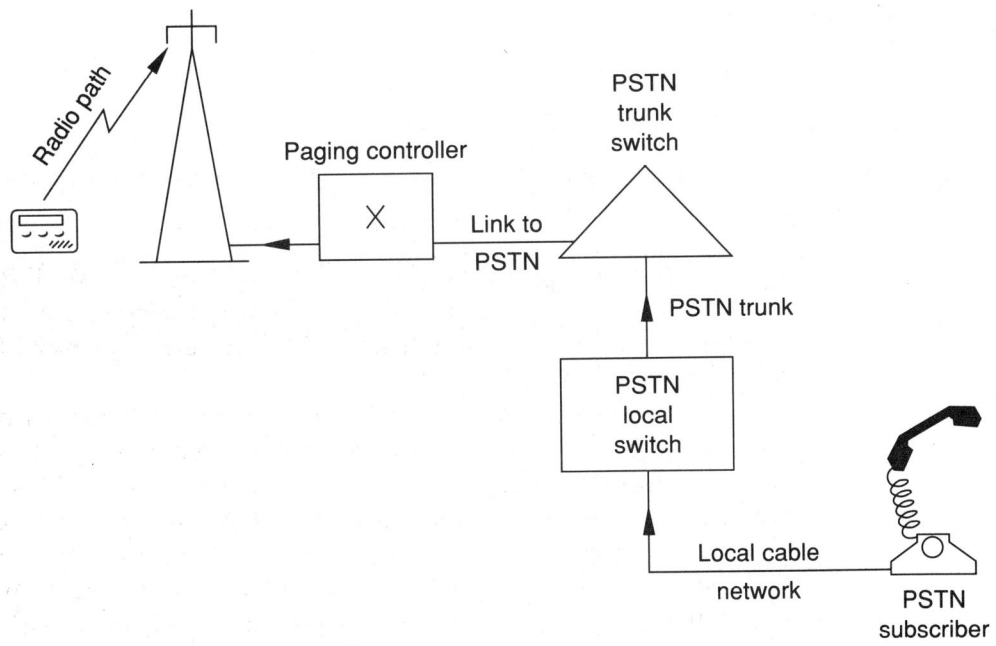

Figure 16.2   *The physical path of a call to a pager from a PSTN subscriber.*

$300,000 to set up making the average cost per subscriber $300. To convert this to an annual cost we may allow for 10 percent depreciation, 10 percent maintenance and operations, and 15 percent return on capital investment to derive an annual return required of 45 percent. If the average subscriber uses 100 minutes of interconnect time per year then the cost of one minute of usage is $300 × 0.45/100 or $1.75 per minute.

Here it must be appreciated that an originated call must involve the PSTN party so that any call involves two subscribers.

The cost of a call carried within the PSTN is well established. The average cost of a new PSTN line is around $2,000 and the PSTN would have a large percentage of depreciated assets, which would bring down the average cost per call.

A major difference between PSTN and paging costs is the economic equipment life, which may be as high as 30 years for some PSTN equipment but which ranges from 7 to 10 years for paging equipment. For this reason the average network book value of the PSTN per subscriber would be between $800 to $1,200, although very new and very old networks may lie outside these limits and costs will be around 10 to 20 cents per minute. Note interconnect times will be much longer than air time for the average call.

DETERMINING FAIR COST FOR INTER-CARRIER TOLLS

International toll charges have long been established on a basis of an internationally agreed TAR (Total Accounting Rate). This rate, which is usually somewhat arbitrary, is the agreed value between two international carriers of one call minute of traffic between them. The proceeds of the TAR are then split, usually 50/50, between the operators on the basis that the originator of the traffic pays. This settlement is illustrated in Figure 16.3.

TARs are often relics of an era when transmission costs were far higher than they are today. This means that some serious anomalies exist. In most cases because of net settlement the actual tariff is almost irrelevant: the net outflow is zero. Some countries do have large net outflows because of immigrant or migrant workers calling home. On the other hand the immigrant sourcing countries often earn large net inflows of capital from this imbalance. In the U.S. in 1989 the net outflow settlement exceeded $US 2 billion.

Notice that the TAR need not bear any direct relationship to the actual toll charges paid by the PSTN subscriber.

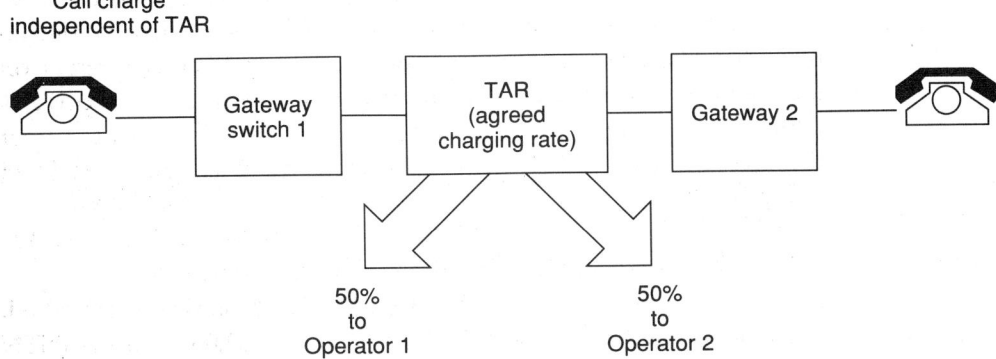

Figure 16.3   *The TAR (assumed here to be $3.00) is usually split equally between the two operators.*

For international calls outgoing from a cellular operator there is thus a clear basis for cost accounting. That is, the country portion of the TAR is the cost. For incoming international calls the cellular operator need only treat them as local PSTN calls (at least for tariff purposes).

National calls may sometimes have an agreed cost value (as for example in the U.S.) but in most countries the only available basis for costing will be the published toll rates. In order to avoid too much debate over the true cost of a national toll call the paging operator will usually negotiate these calls at the published toll rate minus some discount (typically 20 percent to 40 percent).

Based on the above, a formula for settlement between a paging operator and a national PSTN carrier could be as follows:

DUE TO THE PAGING OPERATOR
FOR PSTN-ORIGINATED TRAFFIC

1. A fixed rate per minute of call determined on a cost basis regardless of the origin of the call. For example based on the previous example $0.15 per call minute.

2. Where there is an arrangement between the PSTN and the paging operator that the PSTN operator will collect the air-time charge for PSTN originated calls this will be added.

NOTE   There is an interesting philosophical difference between cases 1 and 2 above. In case 1, it is the paging user who pays. This has an advantage for some users who want to use the pager for business and do not want their customers deterred from ringing in by the cost of making the call. In case 2, the principle that is applied is "user pays," and this would be considered more fair by a lot of operators. An extension of point 1 is that charges are split into a local charge plus a long-distance component.

Other Settlements

A charge is applicable for the use of the circuits that connect the paging operator to the PSTN, and these should be accounted on a usage basis of cost-plus.

The carriers should add a collection fee to their subscribers bill to cover the cost of collections and bad debts.

CHAPTER

17

 CODE FORMAT
AND ERROR
CORRECTION

Analog-system transmissions suffer degradation that is not only inevitable but progressively increases as the number of links in the system increases. Each analog link contributes its own noise and distortion. Digital systems can be constructed so that each repeater, or link, not only amplifies but can regenerate the whole code. Provided sufficient error correction is used, the reconstructed code can have an arbitrarily small error rate.

Paging, being a one-way transmission, can use only forward error correction (FEC). More commonly in two-way communications, automatic repeat requests (ARQs) are used. In the two-way case only error detection is necessary, as an error can prompt a resend. The basic techniques used for error correction and error detection are similar.

BLOCK CODES

Digital data is generally structured into words, which are in turn grouped into blocks. Where error correction code is included in the blocks it is known as block code.

Parity is the simplest form of error detection/correction. In its most basic form, a digital word of, for example, 8 bits might contain one parity bit. This bit is inserted to ensure that the total number of "1s" in the word is either odd or even (depending on the parity selected). In this example the information part of the word is 8 bits. To illustrate how this works, assume that the digital information 1011001 is to be sent.

The total number of "1s" is 4. To construct the word for even parity, the last bit would be a 0, that is the 8-bit word is 10110010, while for odd parity the last bit would be 1 and so the word is 10110011.

In a simple ARQ system this is all that is needed, as once an error is detected then a request to repeat the word can be sent. Notice that this code will identify a single error only and can incorrectly accept a word with two errors if they cause the parity to remain correct. Despite these limitations, this method is sufficient for many high reliability applications.

Extending this concept, it is possible, by using two parity bits per word, to correct actual errors even in the FEQ mode.

To simplify, consider a word of 4 bits total. If each word contains 1 parity bit and each block has one parity word, then the actual bit in error can be detected. For even parity, the word sequence may be

	Information	Parity Bit
Word 1	101	0
Word 2	111	1
Word 3	001	1
Word 4	010	1
Block Parity	001	1

Note that even parity has been used for both the word and block check bits. It would be equally valid to use odd parity for the word check bits and even parity for the block (or vice versa).

Assume now that there was an error in the second bit of word two (that is, the X is a zero instead of a one). We now have

	Information	Parity Bit
Word 1	101	0
Word 2	1X1	1 ----------Error in parity
Word 3	001	1
Word 4	010	1
Block	001	1
Parity	:	
	:	
	:	

error in parity

and it can be seen that the error can be detected at the intersection of the error check column and check row.

Because paging receivers work in a high-noise environment, where an error rate of more than 1 in 16 is common, more sophisticated techniques are needed. One commonly used technique is the Hamming Code.

The block code may consist of any number of message bits, k, together with n-k error-detection bits (where n is the total number of bits in the block). Such code is known as (n, k) code. The dimensionless ratio $r = k/n$ ($0 < r < 1$) is known as the code rate.

CONVOLUTIONAL CODES

Convolutional codes differ from block codes by viewing the data as a continuous message sequence and will generate parity bits continuously with the data flow. This type of encoding is distinguished from block codes by the use of memory, so employed that the encoding is dependent on previously transmitted code.

Convolution codes are relatively low efficiency codes (typically $r = 0.5$) that, because they can correct a continuous string of errors, are ideal for correction when error bursts due to impulse noise are likely to be encountered.

HAMMING CODE

The Hamming Code, devised by Richard Hamming in 1950, is capable of correcting multiple errors. Like most early error correction techniques, the Hamming Code is largely a product of trial and error rather than a systematic and rigorous mathematical approach. The number of redundant bits is determined by the formula

$$2^n >= k + n + 1$$

where

> k = the number of bits in the data word, and
> n = the number of redundant bits.

The code rate for this technique is $r = k/n = 1 - 1/n \times \log_2 (n + 1)$. It is a reasonably efficient code for long code words.

The redundant bits may be placed anywhere in the word, but depending on the transmission mode there may be positions that will improve the noise immunity.

The Hamming bits are the result of the exclusive-or value (XORed) with value with each information bit with a value of "1." At the receiver the Hamming bits are extracted and XORed with all "1s." The result gives the position of single bit errors.

MODULO-2 ARITHMETIC

At the heart of most modern error-correction codes is Modulo-2 arithmetic. In essence it is simply described by the following rules for binary numbers:

a. all digits must be 1 or 0

b. $1 + 1 = 0$

c. $1 + 0 = 0 + 1 = 1$

d. $0 + 0 = 0$

e. addition of binary numbers is accomplished with the above rules plus the rule that there is no carry function.

Addition

For example add 111001 to 100011 as below

```
      111001
   +  100011
      ------
      011010
```

The rules a to d, above, are just those performed by an XOR gate.

Subtraction

Subtraction is identical to addition.

Division

Division is carried out in the same manner as ordinary division. For example, divide 111 into 11001

```
              11
       111  ⌐ 11001
              111
              -----
              00101
                111
               -----
                 10
```

Multiplication

Multiplication is performed the same as conventional multiplication except the addition is Modulo-2. For example, multiply 111 by 110

```
         111
      ×  110
        -----
         000
       111
       111
       -----
       10010
```

The operation of Modulo-2 arithmetic will not produce results that agree with conventional arithmetic but, more importantly, it will produce consistent results that can be reproduced in easily constructed hardware. The most commonly used codes, which are subsets of the cyclic block codes (to be discussed later), were developed in the mid-1950s when the first mathematically derived codes began to appear. Using Modulo-2 arithmetic, the ease of realization is not so much a coincidence but at the time may have been a necessity. Of course today, with microprocessor technology, there are few restrictions on the mathematical complexity of the calculations.

CYCLIC BLOCK CODES

The most commonly used codes today are derivatives of cyclic block codes. A cyclic code is one in which any code word can be shifted end-about to form another code word. Consider the code formed by the message bits m_1, m_2, m_3, m_4 and the code check bits c_1, c_2, c_3 to form the code word

$$m_1 \ m_2 \ m_3 \ m_4 \ c_1 \ c_2 \ c_3$$
$$1 \quad 0 \quad 0 \quad 1 \quad 0 \ 1 \ 1$$

when shifted one place to the left gives the word

$$0 \quad 0 \quad 1 \quad 0 \ 1 \ 1 \ 1$$

where 0011 is a valid message word and 111 the corresponding check sequence. The prominence of cyclic codes is due mainly to the fact that they can be easily implemented by straightforward, Modulo-2 hardware circuits. The word and its corresponding check bits are generated by dividing the message bits (the n bit message being padded with zeros to the total word length, n) by a binary number of length (n – m + 1). This binary number is often referred to as the generator polynomial. The important part of this division is the remainder, which forms the check bits. A characteristic of the word so generated is that it is a perfect multiple of the divisor. So a simple check for error by the receiver is to divide the received word by the divisor and any remainder other than zero indicates an error. With a careful choice of divisor, error correction is also possible.

The next task is to select a suitable divisor. There are no simple rules to selecting a divisor, and the codes have high redundancy. The codes are divided into classes based on the choice of divisors.

BCH Codes

BCH codes (BCH stands for the inventors Bose, Chaudhuri, and Hocquenghem, who derived the code in the 1950s) are the most commonly used codes today. The code is specified by its total and message length as (n, m).

Table 17.1   *BCH codes for word lengths up to 31*

			Divisor								
n	m	t									
7	4	1								1	011
15	11	1								10	011
15	7	2							111	010	001
15	5	3						10	100	110	111
31	26	1						11	101	101	001
31	21	2				1	000	111	110	101	111
31	16	3			101	100	010	011	011	010	101
31	6	7	11	001	011	011	110	101	000	100	111

n = total word length
m = message bits per word
t = maximum number of detectable errors

BCH codes are such that they can be devised to correct any given number of random errors per code word. For block lengths of up to a few hundred bits, they are among the most efficient in terms of total block length and code rate. Table 17.1 gives the BCH divisors for word lengths up to 31.

POLYNOMIAL CODES

The most general class of cyclic codes are known as polynomial codes. A general code word can be described by a polynomial as;

$$f(x) = 1 + x + x^2 + x^3 + \ldots\ldots\ldots + x^{n-1}$$

where x can only take a value of 1 and the power of x signifies its position in the word. It is important to realize that this polynomial is not a conventional one, and for the purposes of this text it can be regarded as a shorthand descriptor of the code bits. The factor x^k, where $x = 0$, is implied for all values of $k < n - 1$ where the kth power is not in the equation.

For example, the CCITT V41 (256,240) code uses the 17-digit divisor

10001000000100001

which could be described by the polynomial

$$g(x) = 1 + x^4 + x^{11} + x^{16}.$$

GOLAY CODE

The Golay code is generated by the divisor, or generating polynomials,

$$g(x) = 1 + x^2 + x^4 + x^5 + x^6 + x^{10} + x^{11}$$

or

$$g(x) = 1 + x + x^5 + x^6 + x^7 + x^9 + x^{11}.$$

The Golay codes can detect any combination of three errors in a 23-bit block.

EARLY FORMATS

The first paging format was two tone sequential. The very first paging receivers used resonating reed tone detectors, and decoding took up to 7 seconds per page. Tone and voice paging were both available.

The 1960s saw the advent of 5-tone sequential formats. These systems were used for both tone and voice paging and, using electronic decoders, could achieve paging rates of 5 pages per second. The system, however, was not suitable for alpha or numeric data.

The next decade saw the development of a number of competing propriety digital formats. Although these were no faster than the 5-tone sequential units, they had the advantage of being adaptable to numeric messages, each of which would take about half a second. The lack of a uniform standard caused a reasonable reluctance on the part of the operators to change to digital. Many of the system manufacturers saw a way out through producing transmitters that were capable of handling a number of different formats in a sequential mode. These transmitters were somewhat complex and generally had substantial teething problems.

A de facto standard came in the late 1970s with the POCSAG code. This code was later accepted as an international standard by the CCIR in February, 1981. POCSAG was the most successful of the third generation codes.

THIRD-GENERATION PAGING CODES

There are three competing third-generation codes widely used in paging systems. NEC has its own code (known as the NEC code), Motorola has Golay, and POCSAG, a British Telecom code, is used by most other manufacturers, probably mainly because it was non-proprietary. The UN body, the Consultative Committee for International Radio (CCIR), has chosen POCSAG as the "international" standard because of its speed, efficient error correction, and multiplicity of manufacturers.

POCSAG (Post Office Code Standardization Advisory Group)

POCSAG was renamed CCIR Radio Paging Code No. 1 (RPC1) when it was adopted as the international standard. The support role for the format has been taken over by CCIR, to the extent that the Post Office Advisory Group has been disbanded. The CCIR recommendation in favor of RPC1 was non-exclusive, and so new international codes potentially may be considered in the future.

Undoubtedly the facts that POCSAG was the code most widely used by manufacturers of pagers (and to a lesser extent paging transmitters—as many of these are optionally available in all formats) and that CCIR has to satisfy the world community strongly weighted the choice of systems in favor of this format.

The error correction code of RPC1 is such that the code with 6 percent of error correction code is about as efficient as 15 percent error code in the Golay format.

POCSAG is widely used for tone paging. It is usually transmitted at a data rate of 512 bits per second and can transmit 15 tone-only calls, five 7-character display, or one 40-character display per second. In the UK and Sweden, 1200 bit/s has been used, and in 1991 there were tests under way on 2400-baud POCSAG. There is no standard speed for this format. The coding is in 32-bit words of 31,21 BCH, which means that it has 2 to the power of 21 combinations, or a code capacity of 2,097,152. The modulation is FSK.

POCSAG FORMAT (also known as RPC No. 1)

The POCSAG signal format consists of a 576-bit synchronization preamble followed by batches of code words.

The preamble consists of the sequence 10101010...... repeated a minimum of 576 times. The preamble precedes each batch of calls. In turn each batch is preceded by a synchronizing code word. Once the current batch is completed, the transmission ceases and subsequent transmissions begin with the preamble, as seen in Figure 17.1.

Each batch consists of the synchronizing code word (SC) followed by eight frames, each of which contains two code words. This structure is shown in Figure 17.2. The SC is itself one code word long, so a frame consists of 17 code words.

The frame structure is such that each of the eight frames (numbered 0–7) is assigned to eight groups of pagers. This means that any individual pager will permanently be assigned to a particular frame and need only "listen" to the address information in its own frame.

The frames consist of two code words, the pager address and the message, plus parity bits, as can be seen in Figure 17.3. Each frame is only 18 bits long, but the actual address of each pager (which is always preceded by a "1") is 21 bits long. However, the first three least significant bits are redundant; they serve merely to identify the frame number that contains the pager's address.

The function bits (see Figure 17.3) are mostly used to allow multiple messages on a single pager, such as different "beep codes." In

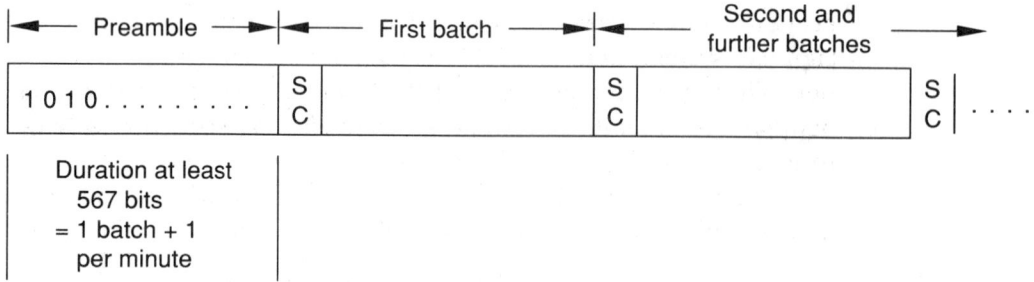

Figure 17.1   *The POCSAG signal format*

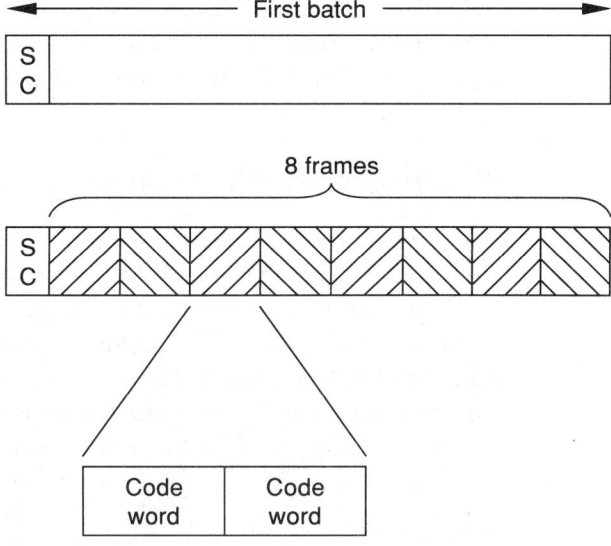

Figure 17.2   *The POCSAG batch structure.*

effect these are an extension of the address code. This two-bit word stands for alert tones, known as A, B, C, and D. The bits are defined as

Bit	Function
00	A
01	B
10	C
11	D

The message bit is only 20 bits long, but this does not limit the message length, because the message code word format can depart from the structure so far described and additional code words can be sent until the message is complete. The arrival of the message flag "0" signifies the next address. The message may even spill over to the next batch, but if it does the batch structure of 17 code words will be maintained. After the message is complete, the sequence will be restored by placing the next message in its correct frame.

When a frame has no message an idle code word is transmitted in lieu of an address. The idle codeword has the following format:

```
BIT No    1 2 3 4 5 6 7 8 9 10 11 12 13 14 15 16
CODE      0 1 1 1 1 0 1 0 1  0  0  0  1  0  0  1
```

BIT No   17 18 19 20 21 22 23 24 25 26 27 28 29 30 31 32
CODE   1   1 0 0   0 0   0 1 1   0 0 1   0 1 1 1

The synchronization code word has the following structure:

BIT No   1 2 3 4 5 6 7 8 9 10 11 12 13 14 15 16
CODE   0 1 1 1 1 1 0 0 1 1 0   1 0 0 1 0

BIT No   17 18 19 20 21 22 23 24 25 26 27 28 29 30 31 32
CODE   0   0 0   1 0 1 0 1 1 1 0 1 1 0 0 0

Pager Call Routing

The pager will usually be assigned a 7-digit telephone number, which need bear no direct relationship to the actual number transmitted. The pagers will have an *internal* identity in the binary equivalent of the numbers 0 000 008 to 2 000 000 (note: 0 000 000 to 0 000 007 are not valid numbers).

The telephone network will transmit the assigned 7-digit dialed number to the paging controller. The controller, in turn, will examine the significant digits of that number and will map them onto the binary number that is actually assigned to the pager. In the example shown in Figure 17.4, it can be seen that the look-up table stores only

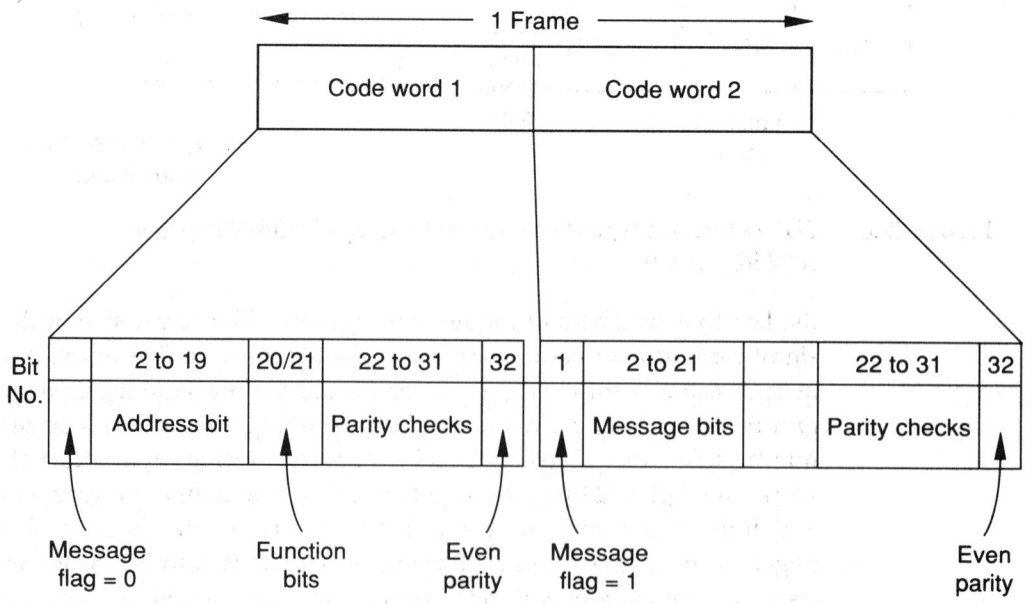

Figure 17.3   *The structure of a frame*

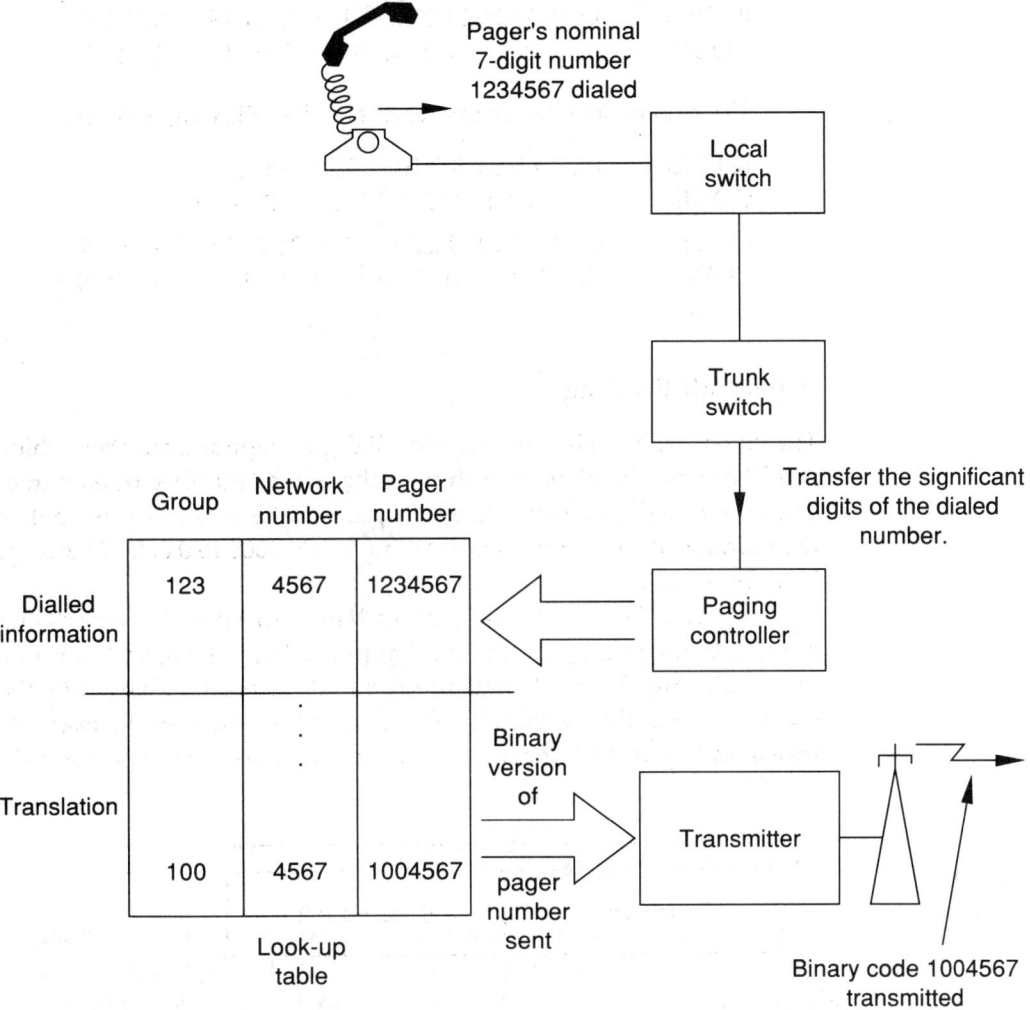

Figure 17.4 *The routing of a typical paging call to pager 1234567 (telephone no.), 1004567 (actual)*

the last four numbers of the network number. The actual size of the significant number code depends on the network configuration, but in this case it is implicit that the prefix 123 simply signifies a pager group (that is, any network number beginning with 123 is a pager number). Because there may be other pager number groups in the network, the digits 123 may be significant if only to define the group to which the pager belongs. Hence, 222 4567 may also be another valid pager number. The actual mapping of telephone number to pager code numbers occurs within the pager controller and will be operator definable. Figure 17.4 shows the path of a typical paging call.

Character	BCD code
0	0000
1	0001
2	0010
3	0011
4	0100
5	0101
6	0110
7	0111
8	1000
9	1001
*	1010
U	1011
space	1100
-	1101
)	1110
(	1111

Figure 17.5   *The numeric character set.*

NUMERIC DATA

In a numeric-only call, the function bit is set to 00 and the message is transmitted as Binary Coded Decimal (BCD). Each character is a 4-bit word and is defined as in Figure 17.5, and a total of 16 characters are defined.

The message code word can carry five characters, and where the message is longer than this it spills over to the next code word, as previously explained. The character "U" is reserved for "urgent."

Alpha-Numeric Characters

The alpha-numeric character set is considerably more complex, as might be expected, and is based on the ISO 646 7-bit code (see Figure 17.6). Because multiples of 7-bit words will not fit evenly into a mes-

Bit No.	7	6	5							0	0	0	0	1	1	1	1
										0	0	1	1	0	0	1	1
										0	1	0	1	0	1	0	1
				4	3	2	1	Column → Row ↓		0	1	2	3	4	5	6	7
				0	0	0	0	0		NUL	TC (DLE)	SP	0		P		p
				0	0	0	1	1		TC (SOH)	DC	!	1	A	Q	a	q
				0	0	1	0	2		TC (STX)	DC	"	2	B	R	b	r
				0	0	1	1	3		TC (ETX)	DC	= (#)	3	C	S	c	s
				0	1	0	0	4		TC (EOT)	DC	$ (□)	4	D	T	d	t
				0	1	0	1	5		TC (ENQ)	TC (NAK)	%	5	E	U	e	u
				0	1	1	0	6		TC (ACK)	TC (SYN)	&	6	F	V	f	v
				0	1	1	1	7		BEL	TC (ETB)	'	7	G	W	g	w
				1	0	0	0	8		FE (BS)	CAN	(	8	H	X	h	x
				1	0	0	1	9		FE (HT)	EM	)	9	I	Y	i	y
				1	0	1	0	10		FE (LF)	SUB	*	:	J	Z	j	z
				1	0	1	1	11		FE (VT)	ESC	+	;	K	[	k	{
				1	1	0	0	12		FE (FF)	IS (FS)	'		L	\	l	|
				1	1	0	1	13		FE (CR)	IS (GS)	–	=	M	]	m	}
				1	1	1	0	14		SO	IS (RS)	.		N	-	n	-
				1	1	1	1	15		SI	IS (US)	/	?	O	-	o	DEL

Figure 17.6   *The ISO 646 Character Set.*

sage code word, the words themselves wrap around to the next code word. Hence, the first code word will contain two whole characters plus 6 bits of the third character. The next code word will begin with the seventh bit of the third code word.

 MODULATION/ DEMODULATION METHODS

Most analog pager systems use FM for speech and FSK (Frequency Shift Keying) for data. Other modulation systems could be used, but this combination gives the best performance when signal-to-noise (S/N) and simpler modulation methods are the main considerations.

Because of the threshold effect (discussed in the next chapter), the mathematics of noise performance above and below the threshold are very different. It is assumed in most radio applications, the transmissions occur above the threshold level (approximately where the S/N of the off-air carrier is 17 dB). Paging systems operate below the threshold, but since the threshold level occurs within a few dB of the pager levels this point is still relevant. Voice pagers will need to operate above the threshold.

RECEIVER PROCESSING GAIN

The baseband reference gain or processing gain (G_B) is defined as the S/N obtainable at the detector output compared to the S/N at the receiver input if the noise is considered to have the bandwidth of the baseband only. Thus, if:

The received power = P_r

Noise/(Hz) = N_d

Modulating signal bandwidth (Hz) = B_m

then:

Equation 18.1

$$SNR_R = \frac{P_r}{N_d \times B_M}$$

where

SNR_R = reference signal-to-noise ratio

The signal-to-noise improvement factor is then:

Equation 18.2

$$G_B = \frac{SNR_A}{SNR_R}$$

where

SNR_A = actual signal-to-noise ratio

This ratio becomes more meaningful when the results for various systems are tabulated, as shown in Table 18.1.

For SSB – SC and DSB – SC, $G_B = 1$, then, as a relative measure of noise performance, the S/N performance of any modulation mode compared to SSB – SC or DSB – SC can be used. The values of G_B for the major systems are shown in Table 18.2, comparing results with other modulation systems.

Clearly, FM is superior in all cases, with the difference between the various systems being a function of their peak deviations. For this reason, FM was chosen for the speech channels.

The data channels generally use FSK, which can yield excellent S/N performance particularly at low data rates. FSK is used fre-

Table 18.1 *Processing gain of different modulation systems*

SYSTEM	$G_B = \dfrac{SNR_A}{SNR_R}$
SSB - SC	1
DSB - SC	1
DSB	$\dfrac{m^2}{1+m^2}$
AM	$\dfrac{m^2}{2+m^2}$
PM	$(A\varnothing)^2$
FM	$\dfrac{3}{2}\beta^2$
Additional gain with FM pre-emphasis	$\dfrac{\pi}{6}\left[\dfrac{W}{f_1}\right]$
Compander	G_B is a function of level and the compression ratio being maximum at low levels of modulation
SSB = single sideband	
DSB = double sideband	
SC = suppressed carrier	
AM = amplitude modulation	
PM = phase modulation	
FM = frequency modulation *and* m = modulation index $0 \leq m \leq 1$	
$A\varnothing$ = maximum phase deviation for PM deviation	
β = modulation index $= \dfrac{\text{deviation}}{\text{audio bandwidth}}$	
W = baseband bandwidth of modulating signal	
f_1 = 3-dB point for pre-emphasis and de-emphasis	

Table 18.2   *The relative noise performance or processing gains of different modulation systems as signal-to-noise ratios*

SYSTEM G_B	OTHER MODES	PAGING D = 4.7 kHz	AMPS D = 12 kHz
SSB	1		
DSB	1		
AM (m = 1)	1/3		
PM*		2.4 (3.8 dB)	16 (12.0 dB)
FM		3.6 (5.6 dB)	24 (13.8 dB)
* PM using same bandwidth.			
All cellular systems are assumed to have 3-kHz baseband.			

quently for signaling in noisy mobile environments. Until recently, techniques for good digital performance over reasonable bandwidths for speech channels had not emerged, and so only analog systems were considered for speech. Today, techniques are available that transmit good-quality speech over bandwidths less than the base bandwidth.

THRESHOLD EFFECT IN FM SYSTEMS

All angle-modulation techniques exhibit a threshold effect; the result of this effect is that a received signal moves from acceptable quality to unacceptable very rapidly as the signal level drops below a certain critical value. FM systems have a processing gain in S/N performance that is a function of their modulation index, which explains why FM is also chosen as the modulation method for high-quality commercial broadcasts. FM stations use 75-kHz deviation, which ensures a very high processing gain. The consequent lower threshold, however, requires a high input signal. A very noticeable improvement in commercial FM-received S/N can be noted with increased input signal level, especially if the receiver is operating in a poor reception area.

At some level, L_{it}, the baseband reference gain, drops off sharply, and within a few dB of input level can drop to less than unity. L_{it} can be shown (by somewhat arduous theory or by practical measurement) to be in the range of 10–20 dB for practical FM systems. This effect is not present in SSB or DSB systems, but it is present in AM detectors that envelope detectors. But because AM systems have pro-

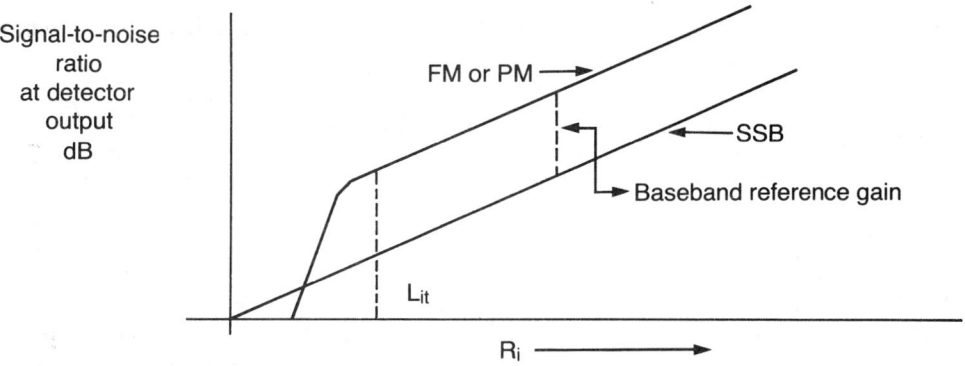

R_i = Receive input signal-to-noise ratio (dB)

Figure 18.1   *Process gain of PM or FM over a linear system such as SSB.*

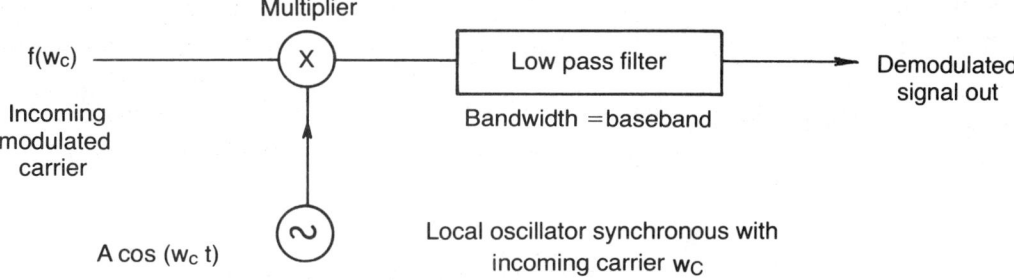

Figure 18.2   *Synchronous product detector.*

cessing gains less than unity, this effect is not so important. SSB systems that use synchronous product detectors do not exhibit this threshold phenomenon, so this mode can be used as a reference.

Figure 18.1 illustrates the threshold effect. Figure 18.2 shows a synchronous product detector.

All types of linear modulation (DSB, AM, SSB, VSB) can be detected by a synchronous product demodulator. For low-level data detection in AM systems, using a synchronous product detector can avoid the threshold effect. There is, however, little practical value in doing this for non-data circuits because the advantages are only realized at very low S/N.

The $f(w_c)$ for an AM signal can be represented as:

$$K \cos w_m t \cdot \cos w_c t$$

where

w_m = modulation frequency

w_c = carrier frequency

If the detector product,

$$P = K \cos w_m t \cdot \cos w_c t \times A \cos (w_c t) = AK \cos^2 (w_c t) \cdot \cos w_m t$$

is taken, we can expand:

$$\left[\text{using } \cos^2 (w_c t) = \frac{\cos (2w_c t) + 1}{2} \right]$$

$$P = \frac{AK}{2} (\cos 2w_c t + 1) + \cos w_m t$$

Equation 18.3

$$P = \frac{AK}{2} (\cos 2w_c t \cdot \cos w_m t + \cos w_m t)$$

Because the term $\cos 2w_c t$ equals the second harmonic of the carrier frequency, a low pass filter can easily remove this product, leaving only:

$$\frac{AK}{2} \cos w_m t$$

(the original modulation).

For a reasonably good-quality signal (acceptable S/N performance), the receiver must operate at an output S/N level of about 35 dB (30-dB S/N is considered the lowest level at which the noise is not obviously intrusive). This is well above threshold.

Because S/N performance determines range, the coverage from the same site of the three main cellular systems can vary significantly. Table 18.3 shows the relative gains of the three systems at S/N output levels of more than 35 dB compared to the lowest, NMT450/900.

BANDWIDTH

The bandwidth required for an FM system is given by Carson's Rule, which states that 98 percent of the power in the sidebands is transmitted if the bandwidth of the system is such that:

Table 18.3 *Relative processing gains at S/N output of more than 35 dB for the three main cellular systems*

SYSTEM	GAIN
NMT450/900	0 dB
TACS	6 dB
AMPS	8 dB

Equation 18.4

$$B_T = 2 \left(\Delta F + f_m \right)$$

where

B_T = bandwidth

ΔF = maximum deviation

f_m = maximum modulation frequency

Table 18.4 shows bandwidth for the systems. Because filters have less than ideal response, the channel filters need to be somewhat wider than BT.

Table 18.4 *Bandwidth versus channel spacing for cellular systems*

SYSTEM	*BT (BANDWIDTH kHz)
4.7 kHz deviation	15.6
12 kHz deviation	30.2
* Assumes f_m = 3.1 kHz (audio speech bandwidth)	

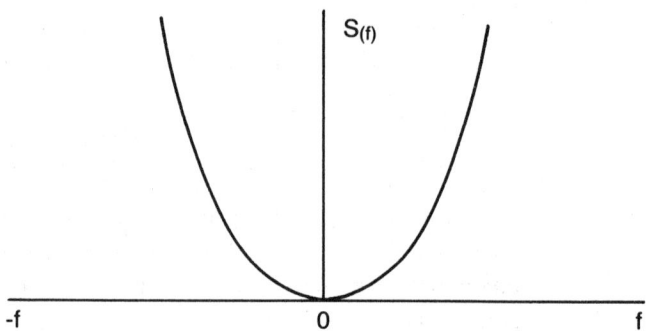

Figure 18.3 *Noise power output of an FM detector as a function of instantaneous deviation.*

PRE-EMPHASIS AND DE-EMPHASIS

The noise at the output of an FM detector has the following density function:

Equation 18.5

$$S_{(f)} = \frac{N_i \times f^2}{2P_r}$$

where

    $S_{(f)}$ = noise density function

    N_i = noise power density at the receiver input

    f = frequency

    P_r = power received

From this equation, it can be seen that the noise power is inversely proportional to the input power, and that this is responsible for the FM quieting effect (that is, the noise level decreases as the input carrier level increases). Figure 18.3 shows the noise power output of FM as a function of deviation. Also, the noise power has a parabolic spectrum as it is proportional to f^2.

The detected S/N, to input S/N can be shown as

Equation 18.6

$$SNR_{(\text{gain})} = \frac{3}{2} \times D^2 \times \left(\frac{B}{W}\right)$$

where

D = peak deviation ratio

B = bandwidth

W = base bandwidth

Thus, the noise power increases rapidly as the bandwidth increases. Fortunately, this can be partially compensated for by pre-emphasis. This involves boosting the transmitted signals in proportion to their frequency.

This is usually achieved using the transfer function

Equation 18.7

$$P_E = K \left(1 + j \frac{f}{f_1} \right)$$

where

P_E = pre–emphasis

f = baseband frequency

f_1 = cutoff frequency

The f_1 is the point at which the modulating signal is boosted by 3 dB. The slope of this curve approaches 6 dB per octave, which is the inverse of the noise spectral density function $S_{(f)}$. In practice, a 6-dB emphasis has been shown to be both readily realizable and capable of producing good results. Increasing the boost has not been found worthwhile. Naturally de-emphasis must be applied at the receive end which has an inverse form:

Equation 18.8

$$D_E = \frac{S_o}{(1 + jf/f_1)}$$

where

D_E = de–emphasis transfer function

f = frequency

f_1 = 3–dB point

S_o = input signal de–emphasis circuit

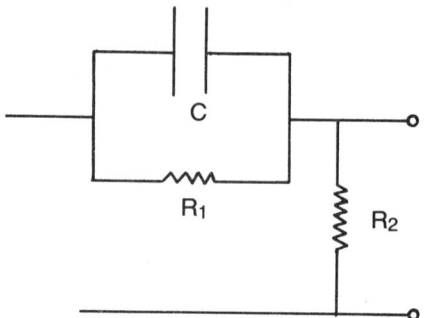

Figure 18.4   *A simple RC pre-emphasis network for FM.*

It can be shown that the improvement in S/N performance due to pre-emphasis and de-emphasis is given approximately by this equation:

Equation 18.9

$$\text{Improvement} \approx \frac{\pi}{6} \times \frac{W}{f_1}$$

where

W = base bandwidth

f_1 = 3–dB point

Nearly all FM systems employ some form of emphasis and de-emphasis, including commercial FM and voice paging systems. A pre-emphasis network can be a simple RC network, as shown in Figure 18.4.

It is normal that $R_1 >> R_2$ so that the time constant of this network is $R_1 \times C$ for the low-frequency cutoff point and $R_2 \times C$ determines the high-frequency cutoff.

Figure 18.5 shows the de-emphasis circuit. This circuit is sometimes described by its time constant $R_1 \times C$ with values of 50, 75, and 100 μsec being common.

SIGNAL-TO-NOISE IMPROVEMENTS WITH A PHASE-LOCKED LOOP

The ultimate performance of an FM system in high-noise conditions is determined by the threshold level as can be seen in Figure 18.1. For

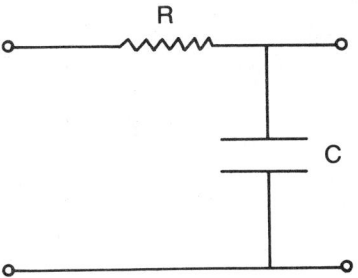

Figure 18.5   *De-emphasis circuit.*

some applications a threshold extension can improve overall performance and range. These techniques have particular application in satellite and deep space communications.

A Phase-Locked Loop (PLL), which can be used in FM receivers, can be designed to improve S/N by 2.5 to 3 dB in the region below the threshold by incorporating a loop response that has spike suppression. Above the threshold, the processing gain of a PLL is the same as a conventional discriminator.

Figure 18.6 shows a simple PLL demodulator. Known as the FM Feedback (FMFB) technique it uses superheterodyne detection. Usually the detection will occur at the IF (intermediate frequency) stage. The difference is that the local oscillator is replaced by a PLL, the

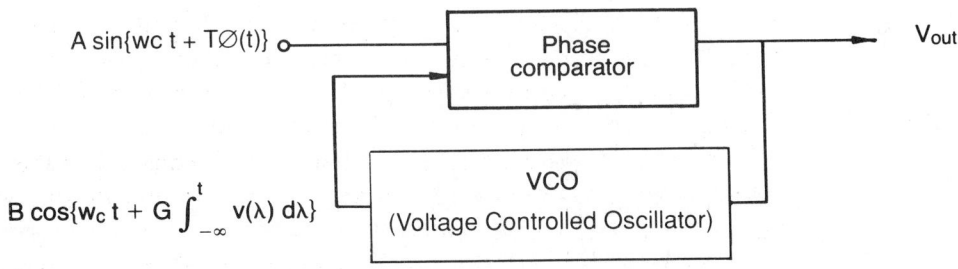

where w_c = carrier frequency

t = time

$\varphi(t)$ = modulation frequency

G = constant

V = output voltage

Figure 18.6   *Simple phase-locked loop detector.*

instantaneous frequency of which is controlled by the frequency of
the modulated signal.

The net effect is to reduce the effective bandwidth of the IF stage
and so reduce the noise power at the output. For the equilibrium of
this loop it is required that

$$\frac{d\phi}{dt} = \frac{d}{dt}\left(G\int_{-\infty}^{t} V(\lambda) \cdot dt\right)$$

and if

$$w = \frac{d\phi}{dt}$$

then

Equation 18.10

$$w = G \cdot V(t) \text{ or } V(t) = \frac{W}{G}$$

so the output voltage is directly proportional to the modulation.

This demodulator is a first-order (or single pole) device; a further improvement can be obtained by using a second-order transfer
function.

Such a PLL can yield an additional 2.5- to 3-dB threshold
improvement. More elaborate PLL detectors are available (with sec-
ond- or third-order transfer functions).

COMPANDING

Companding is the process of compressing the signal before transmission and expanding it at the receiving end. Figure 18.7 illustrates
this process.

A compander, as illustrated in Figure 18.7, reduces the distortion
generated at the transmit end by reducing the voltage excursions of
the modulating signal. It also improves the S/N of an analog linear
system by boosting the level of the low-level signals. The transmission noise (N_t) is added (logarithmically) to the signal on the transmission channel but is reduced by the compander ratio at the receiver.
In FM systems, a compander also improves the S/N performance by
increasing the deviation of the low-level signal components. Figure
18.8 shows a typical transfer characteristic of a compander.

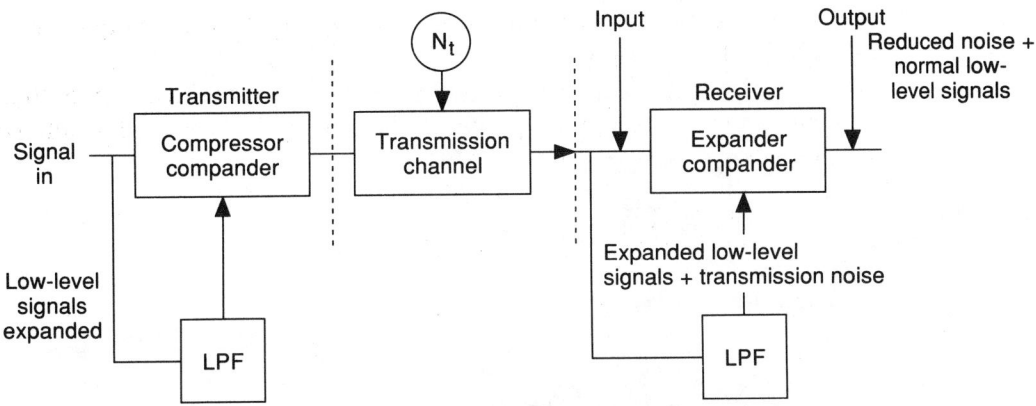

Figure 18.7 *A compander system. The terms "compressor" and "expander" are some-what misleading as the compressor compander compresses signals above the mean level and expands those below it, whereas the expander compander does the opposite.*

This same compression technique is used in some hi-fi tape recorders to reduce the relative S/N and is known commercially as HX noise reduction in some tape recorder systems.

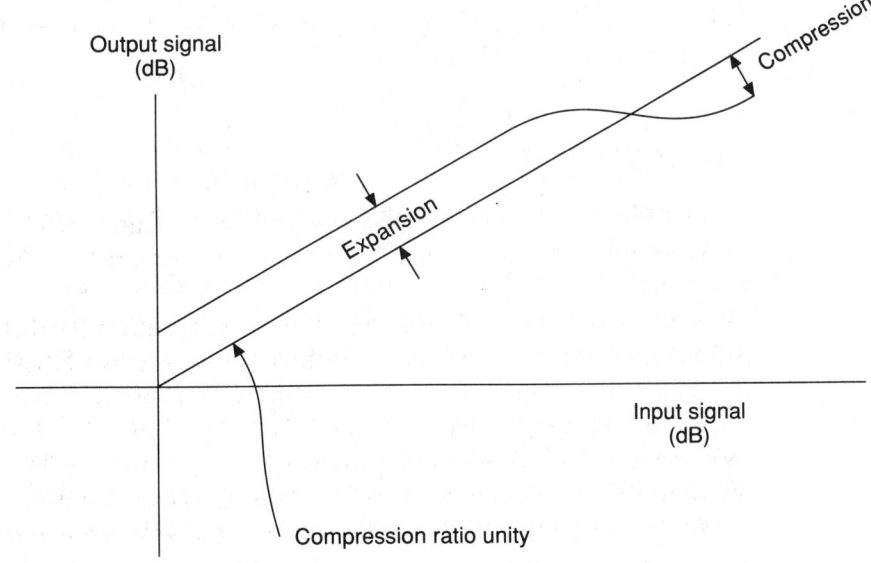

Figure 18.8 *Typical transfer characteristic of a compander.*

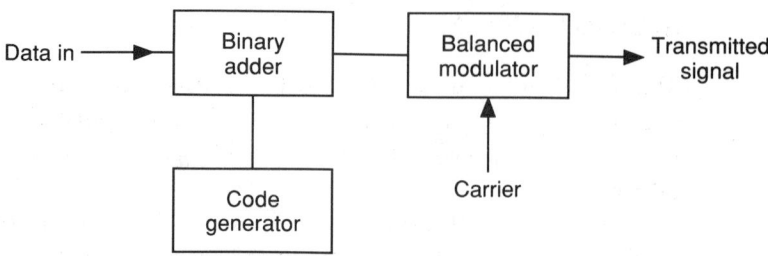

Figure 18.9   *A simple spread-spectrum modulator.*

SPREAD SPECTRUM

Spread-spectrum techniques have long been used by the military because of their high immunity to interference and their high security. It is the interference immunity that particularly appeals to the designers of future cellular systems.

In essence, the spread-spectrum principle is simple. As shown in Figure 18.9, the transmitted digital signal is multiplied by a pseudo-random sequence. The frequency of this sequence is such that it is significantly higher than the information signal and so the resultant modulated waveform "smears" the modulation out over a wide spectral bandwidth. A conventional wide-band receiver would interpret the signal as noise.

In order to decode the signal, the receiver must use the same random sequence as the sender. By multiplying the received code with the decode key the intelligence is removed.

However, if other signals, either wide-band or narrow-band, are received they will not correlate and so will be received as noise. With the use of robust error-correction techniques much of this noise can be filtered out.

Because the pseudo-random sequence is different for each call, many calls can be set up on the same bandwidth and transmitted simultaneously. CDMA is an example of this form of spread spectrum.

Although the military versions of spread spectrum are well understood, the commercial realization of a low cost and relatively low-interference immune system is still a major challenge. Only recently have commercial spread-spectrum ASICs become available.

MODULATION

The usual modulation technique for spread spectrum is phase modulation using either 180 degrees phase-shift or binary phase-shift keying (BPSK). FSK is sometimes used although it does have inferior S/N performance.

When the technique of encoding is a pseudo-random sequence multiplied by the data and then phase modulated it is known as "direct-sequence" spread spectrum (DSSS). An important parameter of direct-sequence modulation is the "chip" or the duration of the smallest bit in the code sequence. This determines not only the spectral bandwidth of the modulated signal but also the processing gain of the receiver, which both increase as the chip duration decreases. The modulating pseudo-random sequence is called the *chip sequence.*

The bulk of the information energy is contained in a bandwidth of $F_0 - F_n$ to $F_0 + F_n$ where F_n = the center frequency of the transmission, and $F_n - 1/(t_c)$, where t_c = chip duration.

The modulator can be a simple exclusive OR gate as depicted in Figure 18.10.

The chip sequence and data stream need not be synchronous although if they are the clock recovery is simplified.

Other techniques include frequency-hopping (as employed in GSM) and time-hopping where the time bursts have a pseudo-random duration.

Another technique known as *chirp modulation,* which consists of linearly swooping over a wide frequency range, is also sometimes classified as spread spectrum, although the lack of a pseudo-random generator means that this is not true spread spectrum.

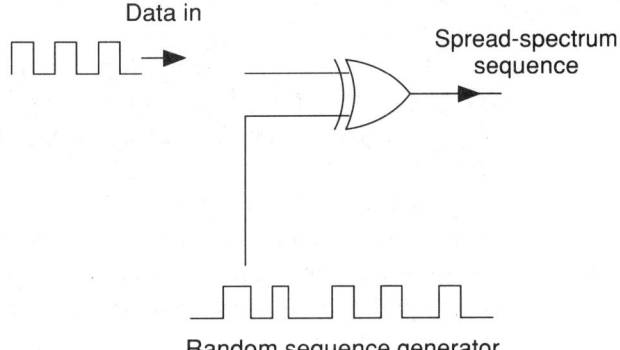

Figure 18.10

CHAPTER
19

 # NOISE AND NOISE PERFORMANCE

All communications systems are noise-limited in the strict mathematical sense that noise determines the maximum data rate that can be transmitted over any fixed bandwidth and with any arbitrary error rate. The bit rate that can be communicated at an arbitrarily low error rate in a Gaussian noise environment was calculated by Claude Shannon in the early 1940s to be

Equation 19b.1

$$C = W \log_2 (1 + S/N)$$

where

C = the baud (or bit) rate

W = the bandwidth in Hz

S/N = signal–to–noise ratio

This equation is true for all signal-to-noise (S/N) ratios (including those < 1), provided a suitable encoding (modulation) system is used.

All analog speech-encoding techniques are very inefficient because speech itself uses very high redundancy and makes poor use of the channel bandwidth.

From Equation 19b.1, it can be seen that the baud (or information) rate is improved with any increase in bandwidth and, for reasonably high S/N ratios, the improvement is almost directly proportional to the log of the S/N ratio. FM was an early and successful attempt to exploit this relationship. In FM broadcasting, a wide bandwidth is used to gain an improved signal quality. Commercial FM broadcasting uses a bandwidth five times broader than the baseband (the modulation frequency).

For HF (High Frequency, 3–30 MHz) the limiting noise is mainly manmade (for example, electrical and ignition systems), but also includes storms, solar flares, and, more importantly, other transmissions on the same frequency (if they are considered to be noise). Thus, with HF, the improvements in receiver technology cannot achieve much improvement in transmission over any single channel. Frequency-hopping techniques, however, can achieve a marked improvement.

Noise can have many origins, including galactic and extra-galactic, thermal, manmade, and even quantum mechanical. Modern radio receivers in the VHF/UHF bands operate close to the theoretical limits of sensitivity imposed by these noise sources.

At VHF and UHF frequencies (30–3000 MHz), and particularly above about 300 MHz (where all cellular radio operates), the background noise is relatively low, and the limits of performance are set by the equipment. Galactic noise can be significant in the region of 40–250 MHz.

At these frequencies, the predominant source of noise is thermal. Modern receivers operate at very low-noise factors (levels of introduced noise) and, short of using very expensive technology (such as masers), few improvements in the basic sensitivity of these receivers are possible.

GALACTIC AND EXTRA-GALACTIC BACKGROUND NOISE

Galactic background noise (which originates within the galaxy) and extra-galactic background noise (which includes the microwave emissions left over from the Big Bang some 16 billion years ago) are significant radio noise sources.

Galactic noise has its origin in stars, supernovas, neutron stars, black holes, quasars, and other noise sources that are scattered throughout the galaxy. Particular noise sources include the sun, Cygnes A (thought to be a black hole that emits high levels of X-rays and radio but only low-intensity light radiation), and Cassiopeia A (a particularly noisy star at radio frequencies). In addition, because the universe is composed of some 10^{12} galaxies, it is not surprising that some of the noise emanates from outside our local galaxy.

Below 1 GHz, the maximum levels of noise are for the beam pointed at the galactic poles. At higher frequencies, the maximum levels are for a beam just above the horizon, and the minimum levels are for the zenith. A low-noise region between 1 and 10 GHz is most amenable to application of special, low-noise antennas.

Antenna noise temperatures are used to define the background noise levels and have nothing to do with the actual temperature of the antenna itself. The noise power is directly proportional to the bandwidth of the receiver. It is important to note that the noise power is independent of antenna gain.

The quietest area is from 1–10 GHz, where galactic noise is at a minimum. Cellular radio systems are usually at 800–900 MHz and operate at antenna noise temperatures of about 2.5 degrees K–100 degrees K. The radio noise is very much a function of frequency, as the Earth's atmosphere acts as an attenuator at high frequencies, while the ionosphere attenuates the lower frequencies. The large variation in antenna temperatures is due to the differences in how the measurements are made. At lower frequencies (below 1 GHz), the maximum temperature occurs when the antenna is pointed at the galactic poles. At higher frequencies, the maximum temperature occurs when the antenna is pointed at the horizon.

THERMAL NOISE

Because of random molecular movement caused by thermal energy, all passive and active components at any temperature above absolute zero generate a certain amount of wide-band energy. In electronic devices, this energy manifests itself as system noise and imposes fundamental limits on usable sensitivity of receiving and detecting systems.

In any conductor the available noise power can be determined by this relationship:

$P = KTB \; watts$

where

P = available noise power

T = temperature in degrees absolute (Kelvin)

K = Boltzman's constant = 1.38×10^{-23} joule/Kelvin

B = the bandwidth in Hz

Notice that the noise power depends only on temperature and bandwidth and is not dependent on resistance.

For a conductor with a resistance R (see Figure 19.1), it can easily be shown that

Equation 19b.2

$$E^2 = 4\,RKTB$$

where

R = the equivalent resistance

N = the equivalent noise voltage generator

Notice that the voltage is represented as E^2. The square of the actual value of voltage (that is, a value proportional to energy) is used

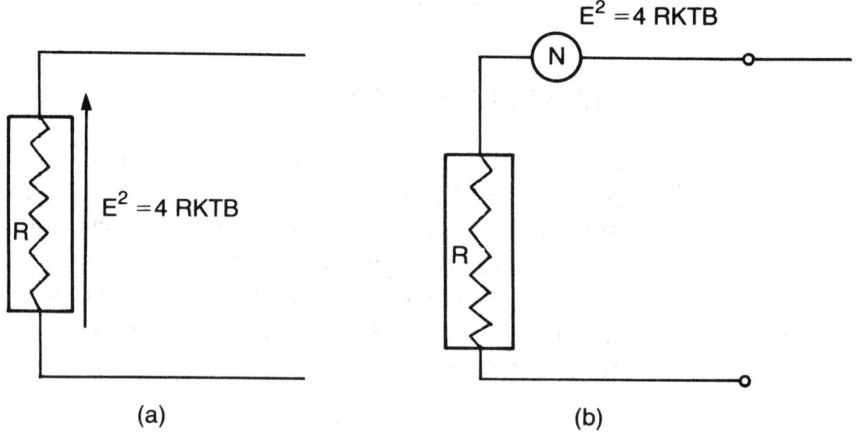

(a)   (b)

R = The equivalent resistance

N = The equivalent noise voltage generator

(a) Voltage across a resistor

(b) Equivalent voltage generating circuit

Figure 19.1   *Equivalent voltage of a resistive noise source, where (a) is the voltage across a resistor and (b) is the equivalent voltage generating circuit.*

in lieu of E, the actual voltage, because the average voltage is zero (that is, noise with negative-going pulses is just as likely to occur as noise with positive-going pulses).

The RMS value of voltage is:

Equation 19b.3

$$E_{RMS} = \sqrt{4\,RKTB}$$

In practice, amplifiers are usually cascaded (that is, they are used in series). The nature of amplifier noise factors dictates that the first stage is the most critical in determining the overall noise performance of a system.

ATMOSPHERIC NOISE

Atmospheric noise is largely due to lightning discharges and is consequently very seasonal. It predominates in the frequency range up to about 20 MHz. Atmospheric noise is not generally a factor at cellular frequencies except in abnormal circumstances.

MANMADE NOISE

Due mainly to low-frequency devices such as motors, neon signs, power lines, and ignition, manmade noise sources tend to decrease rapidly in intensity with increasing frequency. Typically, suburban areas are about 15 dB quieter than city centers, and rural areas are about 15 dB quieter than suburban areas. This noise source is significant up to about 1 GHz, but it is generally not a serious problem above 500 MHz.

SUBJECTIVE EVALUATION OF NOISE

The ultimate determination of S/N performance is the perception of the user. In the audio environment, it is possible to classify S/N in terms of quality. Table 19.1 shows some categories of everyday experience of S/N.

Another method that was developed by the mobile-radio and amateur-radio community is to evaluate S/N on a scale of 1 to 5. Table 19.2 shows this method and its approximate S/N equivalents.

Table 19.1   *Some common S/N levels in everyday systems*

TYPE OF SIGNAL	S/N RATIO (dB)
Limit of operation of 5 tone sequential pager	0–3
Barely readable two-way radio	5–10
Telephone voice quality	25–40
Hi-fi analog recording	55–65
Compact disc	80+

Table 19.2   *Signal quality as a function of S/N ratio*

SIGNAL QUALITY	APPROXIMATE S/N (dB)	SIGNAL NUMBER
Broken and unreadable	5	1
Broken and just readable	10	2
Readable with some difficulty	15	3
Readable with noise	20	4
Clearly readable	25+	5

The fact that this table ends with "clearly readable" is indicative of mobile two-way standards, where a high-quality signal is not generally sought. An S/N of 20 dB would be a low limit of acceptability for cellular subscribers and would be acceptable in fringe areas only.

NOISE FACTOR

In order to look more closely at the noise performance of cellular receivers, it is necessary to introduce the concept of noise factor. Noise factor can be defined as:

Equation 19b.4

$$F = \frac{\text{available S/N power ratio at input}}{\text{available S/N power ratio at output}}$$

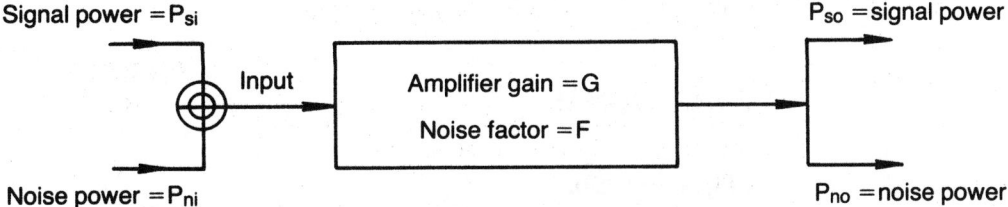

Figure 19.2   *A single-stage amplifier, where the noise contribution of the noise factor F results in the amplifier adding to the output noise.*

where

F = noise factor

Figure 19.2 shows the noise factor.
From Equation 19b.4, you can see that:

Equation 19b.5

$$F = \frac{P_{si}}{P_{ni}} \times \frac{P_{no}}{P_{so}}$$

For linear amplifiers, F is always greater than 1 (that is, noise will be added).

$$G = \frac{P_{so}}{P_{si}}$$

where

G = the gain of the amplifier

Similarly, from Equation 19b.5:

Equation 19b.6

$$F = \frac{P_{no}}{G.P_{ni}}$$

THE AMPLIFIER'S CONTRIBUTION TO NOISE (REFERRED TO THE INPUT LEVEL)

It is often useful to determine the contribution of the amplifier to the overall noise of a system. The output noise referred to the input is P_{no}/G. So, the noise contributed by the amplifier is:

$$\text{AMP Noise} = \frac{P_{no}}{G} - P_{ni}$$

$$= FP_{ni} - P_{ni}$$

Equation 19b.7

Noise contributed by amplifier (referred to input level)
$= P_{ni}(F-1)$

CASCADED AMPLIFIERS

Amplifiers connected in cascade (series) result in an overall noise factor that includes the contributions of each stage. Figure 19.3 shows cascaded amplifiers.

From Equation 19b.9, it can be seen that:

Equivalent noise at the input of amplifier 2

= noise input to amplifier 2 by amplifier 1 plus contribution to noise of amplifier 2 referred to the input (from Equation 19b.9)

$= G_1 \times P_{ni1} \times F_1 + P_{ni2} \times (F_2 - 1)$
(from Equation 19b.9 and Equation 19b.5)

Hence:

Noise output power of amplifier 2

$= G_2 \times$ the total noise input power

Therefore, the noise output power is

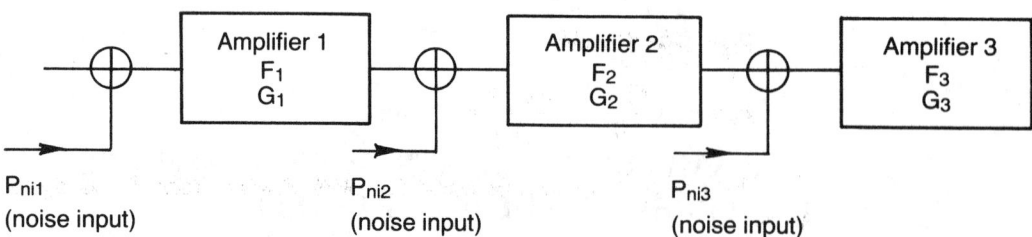

Figure 19.3   *Amplifiers connected in cascade can be regarded as having inputs that are the sum of the previous stage outputs, regarded as NF = 1 stage and the net noise contribution.*

Equation 19b.8

Noise output of stage 2, $P_{no2} = G_2 G_1 \times P_{ni1} F_1 + P_{ni2} (F_2 - 1) G_2$

If the amplifier input and output impedances are matched, and amplifier 2 is at the same temperature as amplifier 1, then:

$P_{ni1} = P_{ni2}$

$= KTB \; watts$

But, from Equation 19b.6:

$$F = \frac{P_{no}}{G \, P_{ni}}$$

which results in the overall gain for series amplifiers 1 and 2 being:

$G = G_1 \times G_2$

From Equation 19b.8:

$P_{no2} = G_2 G_1 \times P_{ni1} F_1 + P_{ni2} (F_2 - 1) G_2$

Then, from Equation 19b.6:

$$F_o = \frac{G_2 G_1 \times P_{ni1} F_1 + P_{ni2} (F_2 - 1) G_2}{G_1 G_2 \, P_{ni1}}$$

But $P_{ni1} = P_{ni2}$ for impedance matched amplifier stages, so the cascaded noise factor F_o is:

Equation 19b.9

$$F_o = F_1 + \frac{F_2 - 1}{G_1}$$

It can similarly be shown that for additional cascaded amplifiers:

Equation 19b.10

$$F_o = F_1 + \frac{F_2 - 1}{G_1} + \frac{F_3 - 1}{G_2 G_1} + \dots$$

Two important conclusions are now drawn from these equations:

■ In the paging receiver environment, the first RF stage virtually determines the noise performance of the receiver. A typical mobile RF stage has the following specification:

Gain = 15 dB (ratio = 31.62)

F = 1.5 dB (ratio = 1.41)

The next stage may have a gain of 30 dB (1000) and a noise factor of 10 dB (ratio 10). So, the overall performance, denoted F_c is:

$$F_c = 1.41 + \frac{10 - 1}{31.62}$$

$$= 1.41 + 0.284$$

$$= 1.694$$

The overall noise factor in dB is 2.28. In other words, the noise factor has not been significantly increased by the addition of a noisy (10 dB) record stage.

It can be further shown that additional amplifiers can be of progressively lower quality (with a higher noise factor and therefore cheaper) without significant degradation in the overall system performance. This fact is most fortunate because the mixer stage in a superheterodyne is very noisy indeed and so, in all high-performance receivers, the mixer is preceded by a low-noise amplifier.

■ The noise factor of a typical UHF receiver can be calculated from S/N ratios using Equation 19b.3, once measurement details are known. Assume the following:

■ S/N ratio = 12 dB
■ Channel bandwidth = 30 kHz
■ Modulation is 1 kHz at 1 kHz deviation
■ Measured sensitivity = 0.2 microvolts
■ Processing gain at these conditions = 5 dB

Then:

$$\text{Actual input noise} = \frac{1}{2} \sqrt{4\ RKTB} \text{ volts } RMS$$

$$= \frac{1}{2}\sqrt{4 \times 50 \times 1.38 \times 10^{-23} \times 290 \times 30{,}000}$$

$$= 0.0774\ V$$

where

> $T = 290$ degrees K or 17 degrees C (an accepted standard room temperature for noise calculations; in practice, this temperature may be somewhat high or low, depending on location)

Hence:

$$K = 1.38 \times 10^{-23}$$

S/N at the input terminals

$$= 20 \log \frac{0.2 + \text{processing gain}}{0.0774} = 9.11 + 5$$

$$= 14.11$$

Thus, the noise factor is 2.11 dB.

Some measurements of signal-to-noise ratio may yield negative noise factors (or less than unity if ratios are used). Negative noise factors can occur when using deviation ratios higher than unity and noise-reduction techniques such as emphasis and companding. The lesson is that S/N ratios mean very little unless the conditions of measurement are clearly stated.

Further, when the measurements are made at very low S/N levels (less than 15 dB at the receiver input), threshold effects can mask the true nature of the S/N performance.

NOISE FIGURE OF AN ATTENUATOR

The mathematics of the noise contribution of an attenuator, such as a transmission line or coaxial cable, are a little complex and involve thermodynamic considerations. Only the result is listed here. Figure 19.4 shows the noise factor contributed by cable loss.

The noise factor of an attenuator F is:

Equation 19b.11

$$F = 1 + (L - 1)\frac{T_c}{T_o}$$

where

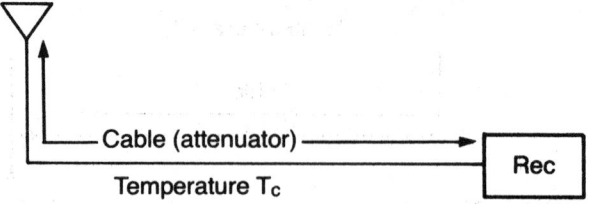

Figure 19.4   *Noise factor contributed by cable loss. The feeder cable to an amplifier is equivalent to a series attenuator.*

L = the attenuation factor

so that

$$L = \frac{\text{power out of attenuator}}{\text{power into attenuator}}$$

$$T_o = 290 \text{ degrees K}$$

$$T_c = \text{cable temperature}$$

When $T_c \approx T_o$, then $F \approx L$ (the attenuator). Hence, the noise factor introduced by the feeder cable will be similar to the attenuation of the cable itself.

As an example, these calculations can be used to determine whether a low-noise mast-head amplifier would be of value in the mobile environment, given the need to operate in a fringe area. Assume the following:

■ The mobile receiver has a noise factor of 6 dB.

■ The cable loss is 3 dB.

■ A mast-head amplifier of 15 dB gain and 4 dB noise factor is available. (This is a low-grade wide-band amplifier.)

■ Receiver gain is 70 dB.

In the case of no mast-head amplifier, the overall noise figure is shown in Figure 19.5.

For $T_c = T_o$:

$$F_{overall} = F_c + \frac{F_r - 1}{G_c} \text{ (from Equation 19b.11)}$$

$$= 1.99 + \frac{3.98 - 1}{0.502}$$

$$= 7.92 \text{ or } 9 \text{ dB}$$

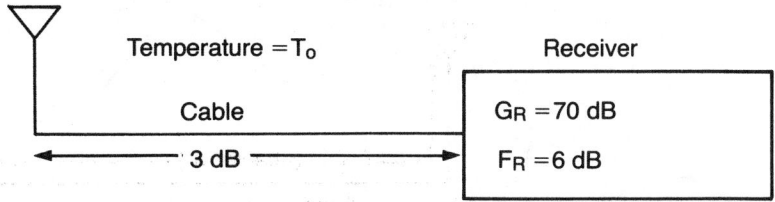

Figure 19.5   *Noise factor of a receiver and feeder cable. The values shown are for a practical receiver with a 3-dB cable and a 70-dB receiver gain with a noise factor
of 6 db. ($G_C = -3$ dB $= 0.502$; $F_C = 3$ dB $= 1.99$; $F_R = 6$ dB $= 3.98$)*

Now, consider the use of a mast-head amplifier, as shown in Figure
19.6:

$$F = 2.51 + \frac{1.99 - 1}{31.6} + \frac{3.98 - 1}{31.6 \times 0.502}$$

$$= 2.51 + 0.031 + 0.187$$

$$= 2.72 \text{ or } 4.35 \text{ dB}$$

So, an improvement of 4.65 dB in S/N performance would result in
this example and, in marginal areas, could well be worth the effort.

It should be noticed here that the "mast-head" amplifier used is
of better quality than the receiver, resulting in an improvement that
exceeds the cable loss. When the amplifiers are of similar quality, (that
is, about a 4-dB noise figure on the first stage of the receiver), the
improvement will be similar to the cable loss, because the contributions to the overall noise factor (F), after the first term, tend to be
small. This system is not necessarily what would result in practice
because mast-head amplifiers with noise factors in the range 0.7–2 dB
are available.

This suggests that mast-head amplifiers are useful for survey
purposes, particularly when low-powered transmitters are used as a
source and measurements are done near the limits of the noise performance of the receiver. Mast-head amplifiers are difficult to install and

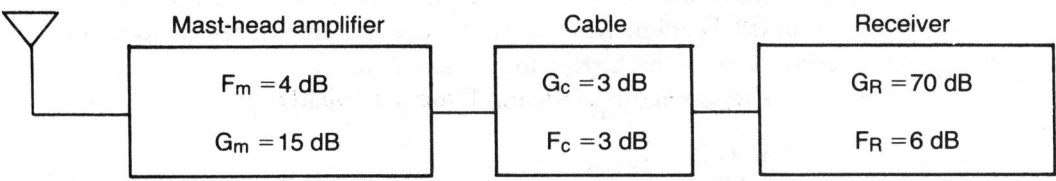

Figure 19.6   *A typical mast-head amplifier is placed as close as possible to the receiving
antenna. ($F_m = 4$ dB $= 2.51$; $G_m = 15$ dB $= 31.6$; $G_C = -3$ dB $= 0.502$; $F_C =
3$ dB $= 1.99$; $G_R = 70$ dB.)*

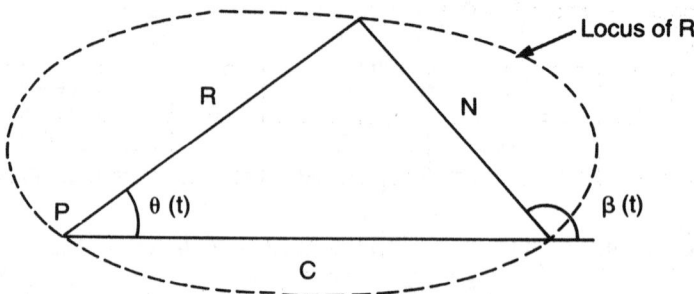

Figure 19.7   *The locus of an FM noise pulse. The received signal is the vector sum of the carrier level (C) and noise level (N). This figure shows the locus of the resultant (received) vector (R), which causes the phase reversal associated with FM "picket fencing."*

are very prone to lightning strikes. In a mobile environment, because they are usually wideband devices, they are somewhat prone to intermodulation, which can limit their utility. They cannot be used on antennas that are connected to a duplexer.

PROCESSING GAIN AND NOISE

The processing gain for an FM system is:

$$G_B = \frac{3}{2}\beta^2$$

where

$$\beta = \Delta f_d / \Delta f_m = \frac{\text{deviation frequency}}{\text{modulation frequency}}$$

However, this formula is valid only at high input S/N levels, where the noise levels are not high enough to cause the noise spikes familiar in FM systems operating in the threshold region. At very high noise levels (at the receiver-input port), noise spikes are generated by the noise components, which are of sufficient level to cause a phase reversal in the incident wave form. Figure 19.7 illustrates the locus of the vector sum of the carrier and noise signals.

    The signal out of the receiver, S_o, is such that:

$$S_o \propto d\theta/dt$$

where

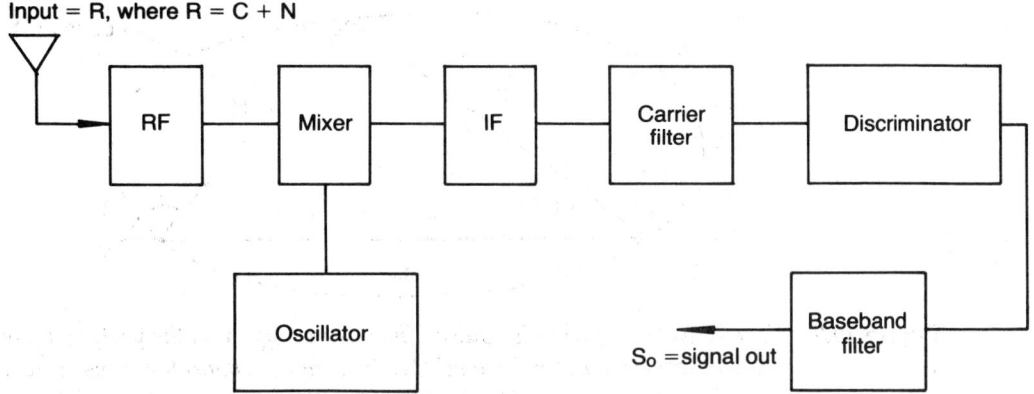

Input = R, where R = C + N

Figure 19.8 *FM receiver block diagram.*

θ = the phase of the resultant of the incident signal and noise

t = time

From Figure 19.7, it can be seen that a phase reversal produces a pulse. If a plot is made of $\beta(t)$, $\theta(t)$, and $d\theta/d(t)$, the form of S_o can be determined. Here, we assume a steady carrier and a noise signal of approximately constant amplitude (but larger than C) that is rotating uniformly with respect to C.

Figure 19.8 shows a typical FM discriminator output filter.

Figure 19.9 shows that this phase reversal produces a pulse. Similarly, it can be shown that when the locus of the resultant does not encircle the point P (as shown in Figure 19.7), a different pulse form arises (as seen in Figure 19.10).

The noise pulse in Figure 19.10 can be shown to be less energetic than the noise pulse in Figure 19.9. Thus, the effect of noise increases rapidly with deterioration of the incident signal. The actual S/N under these conditions can be shown to be:

Equation 19b.12

$$(S/N)_o = \frac{(3/2)\beta^2 (S/N)_i}{1 + (12\,\beta/\pi)(S/N)_i \exp\left[-\frac{1}{2}\{1/(\beta+1)\}(S/N)_i\right]}$$

Note that for large S/N (as encountered in service) the exponential tends to zero and $(S/N)_0 \to 3/2\beta_2(S/N)_i$, as before.

The term in the denominator of Equation L can be seen to tend to 1 when $(S/N)_i$ is large. To determine the levels at which this term

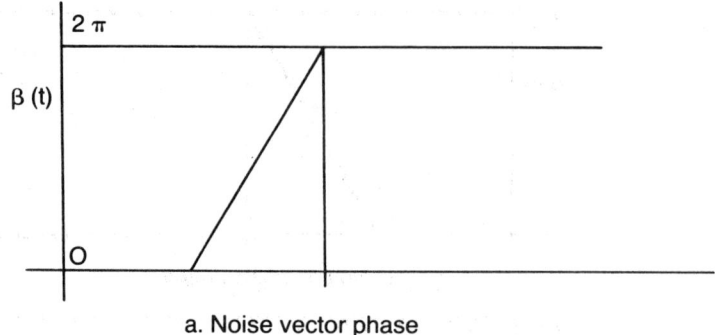

a. Noise vector phase

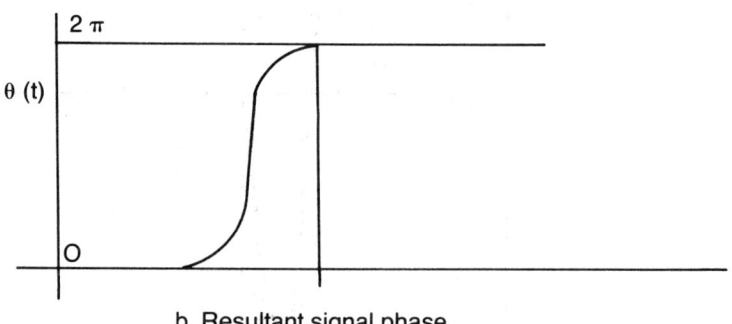

b. Resultant signal phase

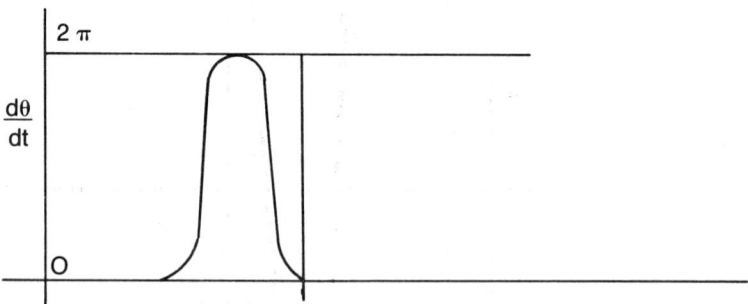

c. dθ/d(t) which is proportional to discriminator output

Figure 19.9   *Noise output pulse generated by a noise pulse that causes the resultant locus to pass through 2π radians.*

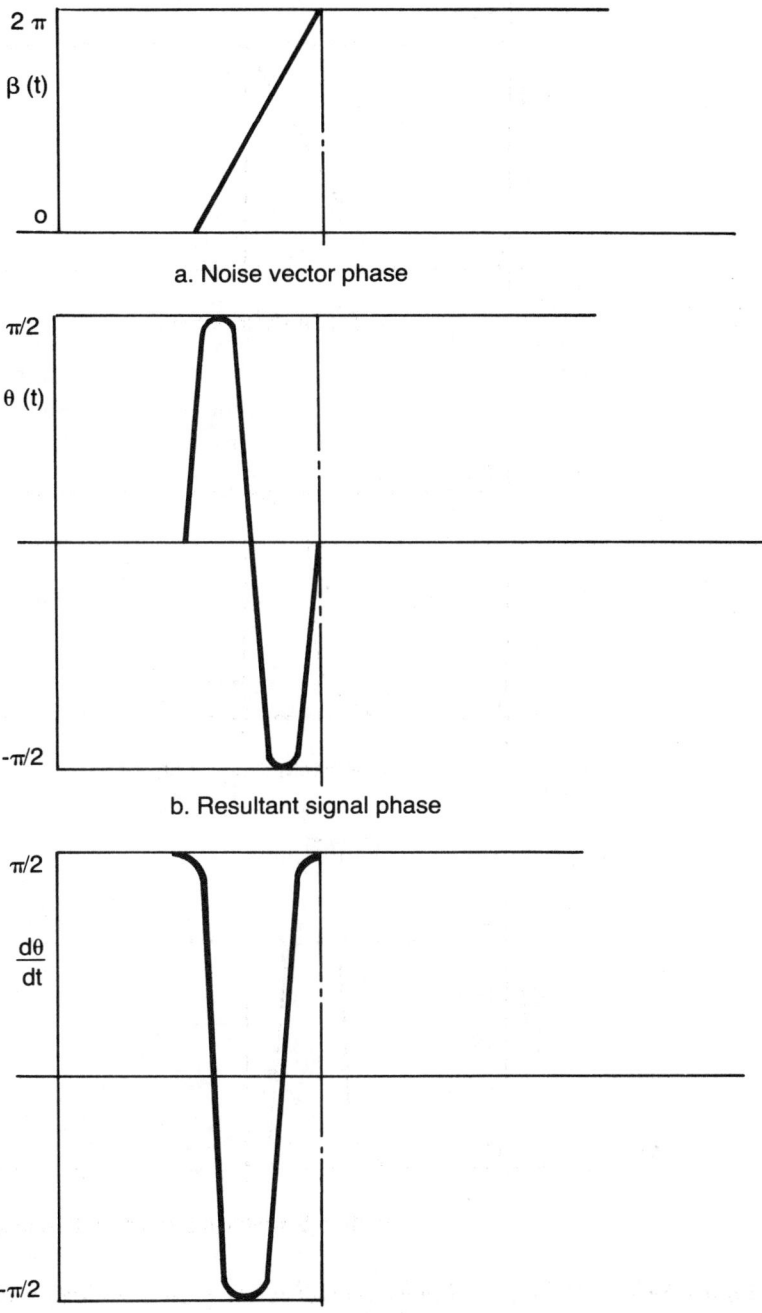

a. Noise vector phase

b. Resultant signal phase

c. dθ/d(t) which is proportional to discriminator output

Figure 19.10 *Noise output pulse from a noise vector N, which is smaller than the carrier and rotates through 2π radians.*

becomes effective, it is traditional to define the threshold as the point where the effect of these terms is to reduce the $(S/N)_o$ by 1 dB, as shown in the following equation:

$$(S/N)_o = \frac{1}{1 + (12\,\beta/\pi)(S/N)_i \exp\left[-1/2\left\{1/(\beta+1)\right\}(S/N)_i\right]}$$

$$= 10^{-0.1} = 0.7943$$

This function can now be plotted for the three major cellular systems (AMPS, TACS, and NMT900). The values for commercial FM are also tabulated. Table 19.3 shows the S/N performance of various systems.

Using Equation 19b.12, the S/N, as measured at the discriminator output of systems with various modulation indexes, can be determined. This function was plotted for commercial FM and AMPS in Figure 19.11.

Table 19.3   *S/N performance of various systems illustrating the relative processing gains*

(S/N)i	COMMERCIAL FM ß=5	AMPS
50	66	63
40	56	53
30	45	43
25	41	38
20	34	33
19	30	31
18	25	28
17	20	24
16	17.3	20
15	14.3	16.3
14	12	13.3
13	10	10.8
12	8.6	8.9
11	7.4	7.4
10	6.5	6.1

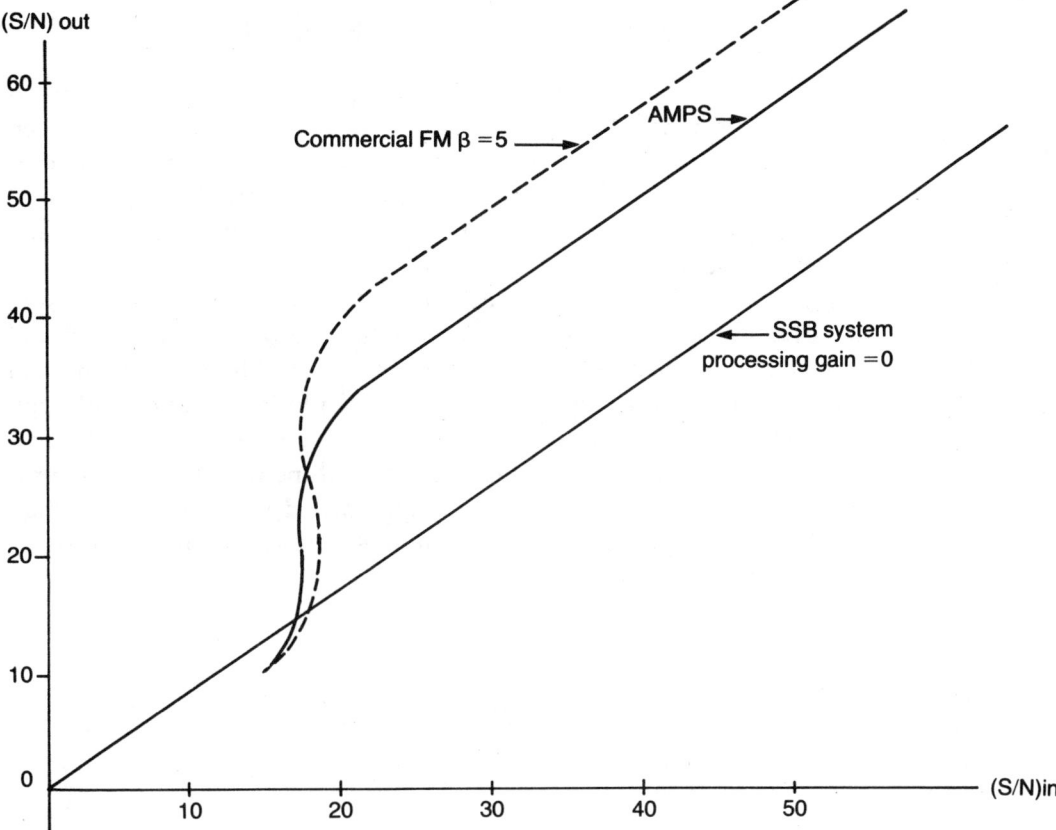

Figure 19.11 *Plots of the processing gains of various FM systems. This graph shows S/N performance versus carrier signal to noise for FM and AMPS systems from Equation 19b.12.*

Above the threshold, all the systems have significant gains over a zero processing-gain system such as SSB. However, at very low levels of $(S/N)_i$ (for example, below 12 dB), the SSB system will out-perform the FM systems.

The various systems have a processing gain that tends to unity in the range 12–13 dB S/N output. For this reason, it is usual to measure S/N at 12 dB SINAD, because this measure reflects the quality of the hardware (noise figure) and will be the same for equal-quality systems regardless of their processing gains.

In all real receivers (as opposed to the theoretical receivers considered previously), the processing gain does not improve the S/N indefinitely, and a threshold is reached where increases in S/N input do not result in increased S/N output.

ABSOLUTE QUANTUM NOISE LIMITS

Without going into extensive theory, quantum mechanics demand that radio noise will exist in a vacuum even at 0 degrees K, even though the vacuum is completely shielded from outside radio influences. In essence, the theory states that a condition of "absolute nothingness," free of any noise, is not achievable, even theoretically, in a perfect vacuum.

Some very sensitive measurements (like those used to measure gravity waves) are now approaching the limits of accuracy permitted by quantum mechanics. These noise effects are different from thermal effects (which are predictable by classical physics), and even though much smaller, they may one day limit the speed of future high-technology data communications by placing a fundamental limit on error rates. Such limits are now being approached in some experimental fiber optic applications. New techniques to reduce quantum noise are being explored.

◆ GLOSSARY

AIR-TIME   The total time that a channel is occupied, including call time, call-set-up, and cleardown time.

ALPHA-NUMERIC   Associated with alpha-numeric systems in which words and numbers are displayed on the pager display screen.

AMPS   Advanced Mobile Phone System.

ANALOG   A system that processes electrical signals by representing the original signal as a continuous function of the original signal. Thus the sound pressure at the input of a microphone may be represented as a voltage level at the terminals of an amplifier, which then amplifies the voltage. Analog differs from digital in that the digital representation of a signal is discrete; that is, all signals are represented as discrete numbers.

ANTENNA GAIN   The gain of an antenna compared to a dipole or quarter-wave antenna. Sometimes the gain is compared to an isotropic antenna, and this is referred to as dBi. dBi = dBd + 2.1.

AREA CODE   Usually a two- or three-digit number that identifies the area of a telephone outside the home area of a caller.

ASIC (Application-Specific Integrated Circuit)   Much like a VLSI except that it has been designed for a particular application only.

Increased use of ASICs lead to smaller size and lower volume cost, however an individual ASIC can be very expensive to develop. The ultimate cellular ASIC is one where the whole cellular radio can be put on one chip.

BASE or BASE STATION   A site that contains the paging transmitting equipment.

BCH CODE   The letters are the initials of the inventors of the format, Bose, Chaudhuri, and Hocqenghem, who developed this structured error correction code in the 1950s. Often described by its total length (l) and message length(m) as (l, m).

BIT ERROR RATE   The number of errors, expressed as a fraction of the total number of bits sent, of a digital signal.

BOUNDARY (OF COVERAGE)   The defined limits of a particular cell. Usually defined to be around 30 dBμV/m for paging systems. However some coverage is usually available well outside the boundary.

CAVITY   A resonant device, usually drum (or cylinder) shaped, that acts as a filter.

CBD   Central Business District (city center).

CDMA (Code Division Multiple Access)   A wide-band spread-spectrum system whereby many RF users can share the same spectrum simultaneously, discriminating the signals by the code that is sent.

CHANNEL   The frequency used by a paging receiver.

COAXIAL CABLE   A pair of conductors consisting of a central conductor surrounded by an outer conductor. These cables are used because of their immunity to interference and relatively low power losses at high frequencies.

COLLINEAR ANTENNA   A gain antenna with dipoles stacked vertically.

COMBINER   A device for combining a number of transmit channels.

CONTROLLER   A switching device located at a central site which controls the paging transmitter.

COUPLER   A device for connecting two or more sources of RF energy to a single cable or port.

COVERAGE   The area over which the service is of an acceptable standard.

CYCLIC CODES   Codes which can be shifted so that when all bits are moved one place (with the last bit replacing the first) a new code word is formed.

dB (DECIBELS)   A unit for expressing the relative intensity of two signals. This is equal to

$$10 \times \log \frac{\text{Power referred to}}{\text{Power of a reference level}}$$

or

$$20 \times \log \frac{\text{Voltage referred to}}{\text{Voltage of a reference level}}$$

dBd   Gain relative to a dipole antenna.

dBi   Gain relative to a hypothetical isotropic antenna. (It is 2.1 times higher than dBd for the same antenna.)

DECODER   A device in the pager that changes the encoded signal into an alert.

DEVIATION   The amount of frequency change from the center frequency in a modulated FM system (expressed in kHz or 1000 Hz).

DIFFRACTION   Propagation around an obstructing object.

DIGITAL   A processing system whereby the processing is done using a number of discrete levels to represent the signal. The simplest digital representation is a binary one; that is, one that has two levels or states—for example, on and off. Any number can be translated into a binary number consisting only of a string of 1s and 0s.

DTMF (Dual Tone Multi Frequency)   The signaling used on modern push-button telephones. Combinations of two tones represent various numbers.

DUAL ADDRESS   Also known as "dual tone" or "duo tone." Two different phone numbers are used to activate their own unique alert. It allows you to know, for instance, whether to call home or the office.

ENCODER   A device that generates signals to a pager.

ERLANG   A unit of telephone traffic such that 1 Erlang is one occupied circuit per hour.

ERP (Effective Radiated Power)   The power, expressed in watts, that is radiated in the direction of maximum antenna gain calculated by multiplying the power at the antenna terminals by that gain. It is often a theoretical power as antenna gains often neglect antenna losses.

FCC (Federal Communications Commission)   The regulatory body in the U.S.

FEEDER   A coaxial cable or waveguide connecting a transmitter/receiver to an antenna.

FEEDLINE   Same as FEEDER.

FM   Frequency Modulation. A very common analog modulation technique noted for its excellent signal-to-noise (S/N) characteristics. The frequency of the carrier varies in proportion to the amplitude of the modulating signal.

FREQUENCY   The rate at which the electric and magnetic fields of a radio wave vibrate per second. Frequency is usually expressed in MHz (1,000,000 Hz); 1 Hz (Hertz) = 1 cycle per second.

FSK (Frequency Shift Keying)   A modulation method using frequency changes (in steps) to transmit data. Usually only two frequencies are used.

GAIN   The factor, usually expressed in decibels (dB), by which the signal received is amplified or improved (in the case of an antenna).

GOLAY   A proprietary paging code often used by Motorola.

GOS (Grade Of Service).   The probability that a call will fail due to the unavailability of links or circuits. Typically paging systems use 0.01 GOS.

GROUND PLANE   The area directly below a quarter-wave or other unbalanced antenna. It should be of low resistance and at least a quarter wavelength in radius from the antenna.

GROUP CALL   A feature found in some paging systems which permits alerting of all units in a given group simultaneously. When used in combination with individual paging, the pager must be capable of responding to two different codes.

HAMMING CODE   An early and somewhat unsystematic code for error correction.

HEXADECIMAL   A number system based on 16.

IC   Integrated Circuits. The building blocks of modern electronic devices.

INTERFERENCE   The reception of unwanted signals that are impressed on the desired signal.

ISOLATOR   A unidirectional RF device, which allows the signal to pass in one direction only.

ISOTROPIC   The same in all directions; in antennas, equal radiation in all directions.

kHz (KILOHERTZ)   One-thousand (kilo) cycles per second; 1000 kHz = 1 MHz.

LEAKY CABLE   A cable that is designed to deliberately leak RF energy. This is often used to provide coverage in tunnels and basements.

MAINTENANCE   Restoring a unit to working order by replacing it or replacing an integral module (for example, a panel).

MAST   A guyed structure meant to support antenna(s).

MEMORY   Integrated circuits that store information such as telephone numbers.

mW (MILLIWATT)   One-thousandth of a watt.

MIS   Management Information System. A software package containing billing and management information.

MODEM   Modulator/demodulator that converts binary to analog signals and analog to binary signals. Used to connect digital devices like computers over analog telephone lines.

MODULATION   The method by which the transmitted signal is impressed on the carrier.

MODULO-2   A form of arithmetic with very simple rules that is in most modern error correction codes.

MTBF   Mean Time Between Failures.

MULTIPATH   The interference patterns created by the addition of signals from more than one path. Virtually all mobile systems operate in a multipath environment; as distinct from point to point systems, which are line of sight and usually there is only one path.

NiCad or NICAD   A nickel-cadmium battery. A rechargeable battery of the type commonly found in cellular handhelds and other mobile handheld radios.

OMNIDIRECTIONAL ANTENNA   An antenna radiating energy equally in all directions (horizontally) around it.

PAGING OPERATOR   The owner and/or operator of a paging network.

PCM (Pulsed Code Modulation)   A digital transmission that uses a number of channels over the same bearer in different timeslots.

PM   Phase Modulation. An analog modulation form related to FM in which the phase of the carrier is varied with the amplitude of the modulating signal.

POCSAG (Post Office Code Standard Advisory Group)
Now renamed Radio Paging Code No 1. (RPC1)—the most commonly used paging code.

POT   Plain Old Telephone.

PTT (Push To Talk)   A radio switch (usually part of the microphone) that must be pushed before the user can transmit. Usual in two-way radio (PMR).

REFRACTION   Propagation other than in a straight line due to bending of the path by some material medium (for example, air or water). Refraction occurs when a change in density of the medium exists.

REPAIR   Restoring a unit to working order by replacing or reconfiguring some internal component, usually at the board or component level. This usually involves bench work with test equipment.

REPEATER   A device that can be either active or passive that receives an incoming signal and relays it on.

RF (Radio Frequencies)   Varies from 10 kHz to 300,000 MHz.

SINAD   Similar to signal-to-noise, but it adds the distortion products (Signal-to-Noise And Distortion) to the noise power.

S/N (SIGNAL-TO-NOISE RATIO)   The power ratio between the received signal source and the noise source.

SPREAD SPECTRUM   A wide-band radio service that can be CDMA, frequency-hopping, or chirp where the carrier frequency is swept with each data burst.

SWITCH   The PSTN telephone switch.

TDMA (Time Division Multiple Access)   A digital (usually radio) system that allows a number of users to use the same system by being dynamically assigned a particular timeslot on request. Often used to describe rural radio telephone systems that use this mode.

TOWER   A self supporting structure intended to hold an antenna(s), as distant from guyed structures.

TRAFFIC   Calls in progress. Measured in Erlangs as one call for one hour equals one Erlang.

TRANSCEIVER   A transmitter and receiver in one unit such as a mobile telephone, or walkie-talkie.

UHF (Ultra High Frequency)   The radio frequency band from 300 to 3000 MHz.

VLSI (Very Large-Scale Integrated circuit)   These are the main component parts of cellular systems in the early 1990s. A VLSI may perform a range of functions but usually of a common type, such as the whole RF section, a logic controller, or a signaling interpreter.

WAVELENGTH   The distance from a point on a radio wave to the same point on the next wave. Paging wavelengths are from 10 to 0.3 meters long.

WIRELINE   Refers to a carrier who also provides fixed telephone services.

INDEX